JN021679

英語で学ぶ
フーリエ解析とその応用

Introduction to Fourier Analysis and Its Applications

佘　錦華・宮本　皓・川田 誠一
【共著】

コロナ社

まえがき

本書は大学に入学して初めてフーリエ解析を学ぶ学生の皆さんに英語でフーリエ解析とその応用を学べるように書いた書籍です。英語の専門用語には対応する日本語の用語を付記することにより学習効果や理解度を高めるようにしています。

フーリエ解析はジョゼフ・フーリエ（1768–1830）が熱伝導の法則を見出し，その現象を表現する方程式を解く方法として考え出した解析手法です。この方法には不連続な関数を含めた任意の周期関数が三角関数の無限級数で表現できることが含まれています。

フーリエはこの理論を「熱の解析的理論」という書籍にまとめて出版しました。数学者からは理論展開の厳密さについてさまざまな指摘を受けましたが，彼の提案した無限級数の収束性については後の数学者が証明しています。このことがその後の数学の発展に多大な影響を与えました。数学者が書くフーリエ解析の書籍が多い理由はここにあります。

フーリエの元の目的であった熱伝導の解析は，そののちさまざまな物質，さまざまな形状をもつ物体の熱伝導の解析に展開され実用化されています。例えば，建築設計に使われて私たちの生活空間を快適にするのに役立てたり，ロケットや飛行機，コンピュータなどさまざまな人工物で発生する熱を効果的に除去したりするために使われたりしています。

ここで視点を変えてディジタル化が進んでいる今の時代におけるフーリエ解析を考えてみましょう。この本が想定する読者にとって最も役立つ分野です。そしてその理論的基盤はフーリエ変換と逆フーリエ変換です。キーワードはいくつかあり，信号解析，スペクトル解析，高速フーリエ変換，制御工学，線形システム理論などです。これらの分野を学ぶ基礎としてフーリエ解析は必須です。

さて，工学部の多くの学生にとってフーリエ解析は難しいと思われています。新しい概念を理解するまでにたくさんの数式が並んでいるからだと思います。

しかし，基本は足し算，引き算，掛け算，割り算です。そして皆さんがなじんでいる多項式関数，三角関数，指数関数の性質をもう一度確認しましょう。そしてペンとノートを用意し，自分の頭と手で計算して例題にある周期関数をフーリエ級数に展開してみましょう。これができればフーリエ級数はすぐに理解できますし，応用することもできるようになります。

フーリエ変換も同じように学んでください。定義式を用いて例題にある関数のフーリエ変換と逆フーリエ変換を計算し，それを通して複雑な数式に慣れてください。そうすることで本書の後半まで臆することなく読み進めていく解析力が身につくと思います。

本書を手にされた皆さんが本書を読破し，さらなる高みを目指して学んでいくことを切に願っています。

最後に，日本大学生産工学部機械工学科の綱島均 教授に鉄道状況診断データの使用許可をいただき，また，株式会社 CrowLab および宇都宮大学バイオサイエンス教育研究センターの塚原直樹 博士にカラスの鳴き声データをわざわざ拙作のために作成していただき，この場を借りて感謝いたします。なお，例題の作成に協力した東京工科大学大学院の劉鉄城 氏と，原稿を詳しくチェックし貴重な助言をくださった中国湖南工業大学の何静 氏，および中国地質大学（武漢）の梅啓程 氏，羅望 氏，孫一仆 氏，周宇健 氏と賀文朋 氏にお礼を申し上げます。また，出版にあたり心暖かく見守ってくださったコロナ社に深く感謝いたします。

2023 年 3 月

余　錦華，宮本　皓，川田 誠一

Preface

This book is written for students who learn Fourier analysis for the first time after entering university so that they can learn Fourier analysis and its applications in English. To enhance the learning effectiveness and comprehension of students studying at universities in Japan, we add the corresponding Japanese terms to English technical terms.

Fourier analysis is an analysis technique devised by Joseph Fourier (1768–1830) as a method of finding the law of heat conduction and solving the equation that describes physical phenomena. This method includes the fact that any periodic function, including discontinuous functions, can be represented by an infinite series of trigonometric functions.

Fourier published his theory in a book entitled *The Analytic Theory of Heat*. Mathematicians pointed out the rigor of theory development. Later, mathematicians proved the convergence of the infinite series he proposed. This had an enormous impact on the subsequent development of mathematics. This is why there are a huge number of books on Fourier analysis written by mathematicians.

The analysis of heat conduction, which was Fourier's original purpose, was later developed into the analysis of heat conduction for various substances and with various shapes and put into practical use. For example, it is used in architectural design to make our living spaces comfortable and is used to effectively remove the heat generated by various man-made objects such as rockets, airplanes, and computers.

Let us change our perspective and consider Fourier analysis in this age of digitalization. This is the area where Fourier analysis is the most useful field for the readers of this book. The theoretical basis is Fourier and inverse

Fourier transforms. Take signal analysis, spectrum analysis, fast Fourier transform, control engineering, linear system theory, and other keywords into consideration. Fourier analysis is essential as a basis for studying these fields.

Fourier analysis is considered to be a difficult subject for engineering students because there are many formulas involved in the process of understanding a new concept. Nevertheless, the basics are addition, subtraction, multiplication, and division. It is also important to reconfirm the properties of polynomial functions, trigonometric functions, and exponential functions that we are all familiar with. It is strongly recommended to prepare a pen and a notebook, and expand periodic functions to a Fourier series in examples using your head and hands. If you can do this, you will be able to quickly understand the Fourier series and apply it.

Learn the Fourier transform in the same way. Computing Fourier and inverse Fourier transforms of functions examples from definitions in this book ensures getting used to complex formulas. By doing so, a reader will acquire analytical skills to read the latter half of this book without hesitation.

We sincerely hope that everyone who picks up this book will read through it and learn to take it to a higher level.

Finally, we would like to take this opportunity to thank Professor Hitoshi Tsunashima of the Department of Mechanical Engineering, College of Industrial Technology, Nihon University for permission to use the railroad diagnostic data, and Dr. Naoki Tsukahara of CrowLab Inc. and the Center for Bioscience Education and Research and Education, Utsunomiya University for creating crow-call data for this book. We would like to express heartfelt appreciation to Mr. Tiecheng Liu of the Graduate School of Tokyo University of Technology for his help in preparing examples, to Prof. Jing

He of Hunan University of Technology, Zhuzhou, China, and Mr. Qicheng Mei, Mr. Wang Luo, Mr. Yipu Sun, Mr. Yujian Zhou, and Mr. Wenpeng He of China University of Geosciences, Wuhan, China for their careful checking of the manuscript and for her valuable advice. We are deeply grateful to Corona Publishing Co. Ltd. for warmly watching over the publication.

March 2023

Jinhua She, Kou Miyamoto, Seiichi Kawata

Contents

1 Overview of Fourier Analysis

2 Mathematical Fundamentals for Fourier Analysis

3 Fourier Series

4 Fourier Transform

5 Signal Sampling and Reconstruction

6 Discrete Fourier Transform and Fast Fourier Transform

目　　　次

1 Overview of Fourier Analysis

Nowadays, *computed tomography* (CT, コンピュータ断層撮影) scans play an important role in the medical field. It provides us with visual information about the inside of a body to make it easy to diagnose diseases. This requires a technology called the *Fourier transform* (フーリエ変換) to scan a human body with radiation and construct internal images of the body.

Just like listening to music and writing it down in pitch strengths, the Fourier transform of an original function clearly reveals its characteristics in a special domain [it is called the *frequency domain* (周波数領域)] that cannot be seen in the *time domain* (時間領域). Since many physical and engineering phenomena can easily be analyzed using the Fourier transform, as a mathematical tool of applied analysis, it is of† central importance in signal processing and system analysis in physical science, applied mathematics, and engineering.

1.1 History of Fourier Analysis

Jean-Baptiste Joseph Fourier (March 21, 1768–May 16, 1830) was a French mathematician and physicist. He showed that heat conduction in solid bodies could be analyzed in terms of an infinite mathematical series, the Fourier series, in *Théorie analytique de la chaleur* (*The Analytical Theory of Heat*) in 1822. This is the beginning of the *Fourier analysis* (フーリエ解析).

The basic idea is that a periodic function can be represented by a *Fourier*

† be of ～：～という特徴をもつ（= have）。

series (フーリエ級数) (a superposition of simple sinusoidal waves) and an *aperiodic function* (非周期関数) can be represented by a *Fourier integral* (フーリエ積分). If a function depends on time (for example, a sound) or space (for example, a picture), then we decompose it into a function in *temporal frequency* (時間周波数) or *spatial frequency* (空間周波数) to extract its properties. This is called a Fourier transform, which is the cornerstone of the modern digital age. Almost all of the signals around us can be transformed into processable signals using the Fourier analysis.

In science and engineering, it is important to mathematically express and analyze phenomena and derive methods to clarify their essence. The Fourier analysis provides us with a tool for such a purpose. The analysis and its philosophy have an essential influence on science and engineering. Familiarity with this analysis will be of great help in learning specialized subjects.

1.2 Illustrative Examples

This section presents three examples among others[†] to show how the Fourier analysis is used to solve problems.

1.2.1 Shape of Sound

There are two types of sounds: a pure tone and a compound sound. A pure tone is a sound with a sinusoidal waveform, or in other words, a sine wave of any frequency and amplitude. The sound we usually hear is a compound sound, which is a sound composed of several sinusoidal waveforms superimposed on the main one. In **Figure 1.1**, a compound sound is decomposed into three pure tones.

† among others：数ある～の中で。

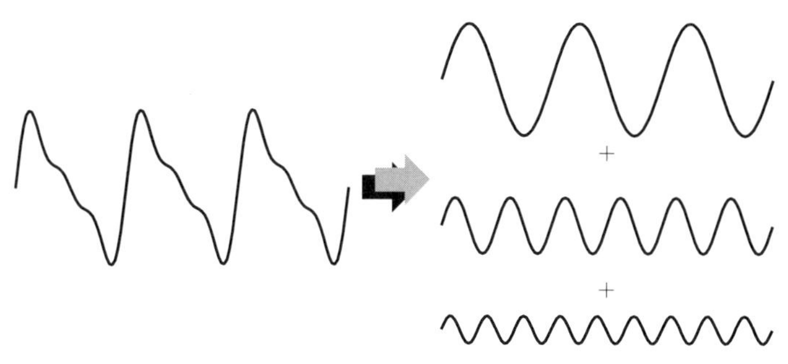

Figure 1.1 Decomposition of a compound sound into three pure tones

Examining the frequency components of pure tones in a compound sound lays the foundation for speech analysis and is widely used in speech recognition, speech synthesis, medical studies including vocal loading, audio forensics, and so on.

1.2.2 Image Processing

While a sound is a function of time, an image is a function of space. Thus, an image is a two-dimensional (height and width) function. In the previous example, the frequency of a sound is the number of occurrences of a repeating event per unit of time because a sound is a function of time. In contrast[†1], if we define a spatial frequency to be the number of striped patterns per unit length, we can analyze images in the same way as dealing with sounds.

An example is shown in **Figure 1.2**(b), in which the central part shows low-frequency components; and the outer part, high-frequency ones[†2]. Observing an image in the frequency domain provides us with a quite different viewpoint. This makes it easy to extract the nature hidden behind complex changes.

---

†1 in contrast：これに対して，それに対して。

†2 “,” の使用で同じ動詞の繰り返しを避ける：このコンマは “shows” を意味する。

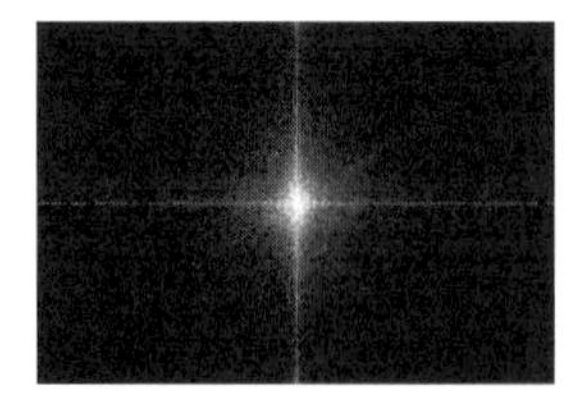

(a) Original image

(b) 2-dimensional Fourier transformation of (a) (Low-frequency components at center)

(c) 2-dimensional inverse Fourier transformation of (b)

Figure 1.2 Fourier transform of a two-dimensional image

1.2.3 Health Monitoring of Railroad Tracks

Track safety monitoring and management are important. Since damage such as rail breakage may lead to serious accidents such as derailment, it is necessary to detect signs of cracks and other damage at an early stage before rail damage occurs to prevent accidents. Equipping an ordinary train with some simple sensors makes it possible to diagnose the condition of tracks while the train is in commercial operation.

Train tracks gradually change due to the repeated passage of trains and natural phenomena. The shape of the rail, which is the running road surface of a train, changes in the longitudinal direction. This is called track irregularity. Rail corrugation is a phenomenon in which the top of the railhead wears from several centimeters to a dozen centimeters. Large rail corrugation causes loud noise and vibrations, and damages to rail material. Note that track irregularity causes big vibrations in low frequencies; and rail corrugation, in high frequencies. It is possible to detect track conditions from noise in a train. **Figure 1.3**[16),†] shows a big peak observed at a low frequency caused by rail corrugation.

† 肩付き数字は巻末の文献番号を示す。

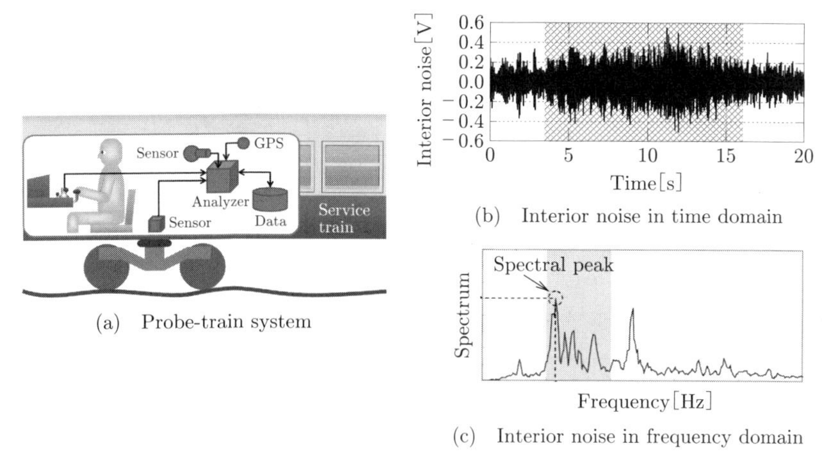

(a) Probe-train system

(b) Interior noise in time domain

(c) Interior noise in frequency domain

Figure 1.3 Probe-train system and monitoring results

1.2.4 Structural Control

The climate and topography of Japan make it particularly vulnerable to natural disasters. Japan is located in the Ring of Fire (also known as the Circum-Pacific Mobile Belt) where seismic and volcanic activities occur constantly. Although the country covers only 0.25% of the land area on the planet, 18.5% of earthquakes in the world occur in Japan, which is an extremely high number.

About 30 typhoons originate over the Northwest Pacific Ocean every year. Okinawa lies right in the heart of Typhoon Alley. Seven or eight typhoons every year pass over Okinawa, and about three hit the Japanese main islands, especially Kyushu and Shikoku. They have hurricane-strength winds, sometimes up to 300 km/h.

How to protect structures from earthquakes, typhoons, and other types of disasters is a big issue. Many advanced theorems and technologies have been applied to deal with them.

Now, we take an earthquake as an example. Different seismic waves

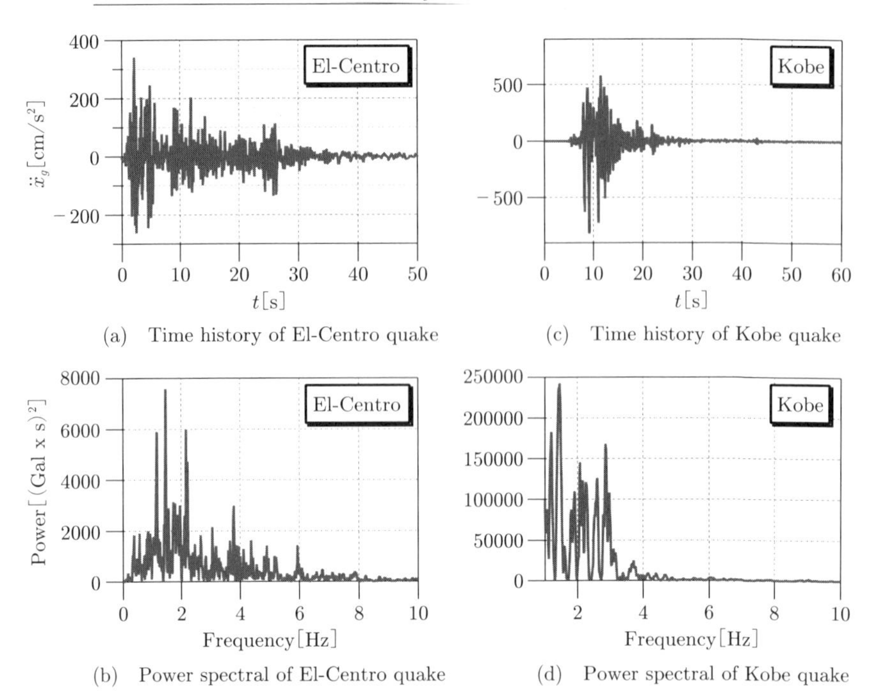

(a) Time history of El-Centro quake

(c) Time history of Kobe quake

(b) Power spectral of El-Centro quake

(d) Power spectral of Kobe quake

Figure 1.4 Seismic waves and their spectra

contain different frequency components (**Figure 1.4**). In order to† suppress structural vibrations, a tuned mass damper (TMD) mounted on a structure is used to reduce the response of the structure to various earthquakes. The key to seismic control is to find a suitable set of the stiffness and the damping coefficient of a tuned mass damper that ensure low sensitivity of the structure during the range of all possible earthquakes. A method called frequency response is used in control engineering to find such a parameter set.

As shown in an example in **Figure 1.5**, while the maximum displacement of the first story without a tuned mass damper is more than 0.4 times a ground motion at a frequency of 1 Hz, that with a tuned mass damper is

† In order to：をするために，をするには。

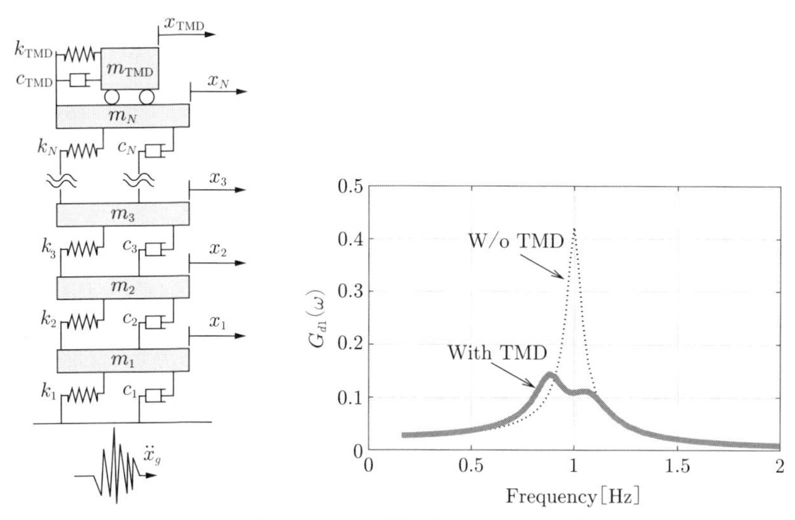

(a) Structure with a tuned mass damper

(b) Frequency gain from a ground motion to the first story displacement

Figure 1.5 Frequency gain from a ground motion to a story displacement

reduced to 0.15 times the ground motion. This shows the effectiveness of the use of the tuned mass damper.

1.3 Key Points of Fourier Analysis

As shown in the examples in† Section 1.2, Fourier analysis finds frequency components of a time or spacial signal (**Figure 1.6**). It is important to understand

(1) what we can find by expressing a function as a sum of trigonometric functions (a Fourier series) or by multiplying a function by a trigonometric function and integrating it (a Fourier transform),

(2) how to properly use a Fourier series and the Fourier transform.

† as shown in the examples in ～：～の例に示されているように。

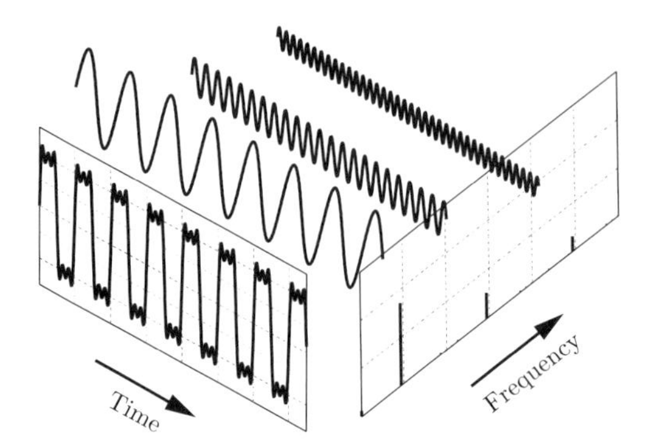

Figure 1.6 Observation of time function

Knowledge of trigonometric properties and operations is the basis of the Fourier analysis. Complex calculations can be carried out using MATLAB/ Simulink[†1, 11)], Scilab[†2, 5), 37)], and other numeric computing programs.

Convolution is widely used in spectrum analysis, control-system design, and many other fields. It is desirable to understand this concept in-depth if there is sufficient time.

This book explains the idea of the Fourier analysis in an easy-to-understand manner[†3] so as to help readers naturally acquire the basic idea. Simple problems are used to deepen the understanding and MATLAB commands for numerical examples are provided to ease learning.

†1 They offer low-price licenses for students and for home use: https://jp.mathworks.com/

†2 A free, open-source software package: https://www.scilab.org/

†3 an easy-to-understand manner：わかりやすい（理解しやすい・把握しやすい）形（表現）で。

Problems

Basic Level

【1】 Find the following technical terms in English:

A. フーリエ級数. B. フーリエ変換. C. フーリエ解析.

D. 周波数応答. E. 周波数特性. F. 振幅. G. 位相.

【2】 Give as many examples as possible to which the Fourier analysis seems to be applicable and briefly explain them.

Advanced Level

【1】 JPEG is a common method that uses the Fourier transform to compress digital images. Explain the compression procedure.

【2】 Explain the basic principles of a computed tomography (CT) (formerly known as computed axial tomography or CAT) scan.

2 Mathematical Fundamentals for Fourier Analysis

This chapter explains some basic mathematics required for Fourier analysis. Refer to[†] college textbooks for details.

2.1 Complex Number

A *complex number* (複素数) is indicated by

$$z = a + jb \text{ [Orthogonal form, } \textbf{Figure 2.1}\text{(a)]}, \tag{2.1}$$

$$= re^{j\theta} \quad \text{[Polar form, Figure 2.1(b)]}, \tag{2.2}$$

where a, b, r, and θ are real numbers and j $(= \sqrt{-1})$ is the imaginary

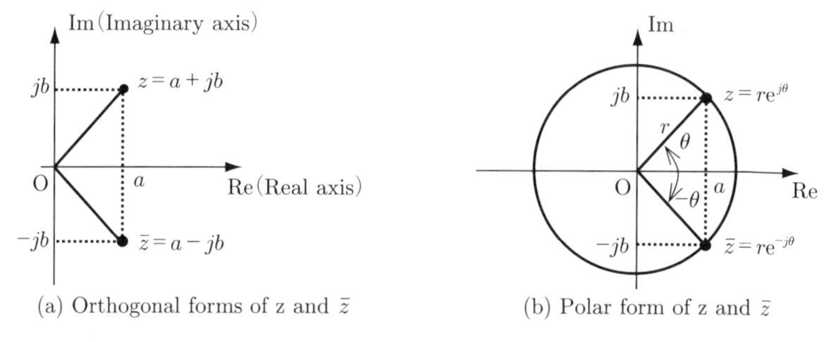

(a) Orthogonal forms of z and $\bar{z}$ (b) Polar form of z and $\bar{z}$

Figure 2.1 Complex number and its conjugate

† refer to：～に参照する。

unit[†1]. a and b in (2.1) are the real and imaginary parts of z, respectively; and r and θ in (2.2) are the distance of the point z to the origin and the angle of the line Oz with the positive real axis in the complex plane, respectively (Figure 2.1).

Euler's formula (オイラーの公式) provides us with

$$\mathrm{e}^{j\theta} = \cos\theta + j\sin\theta, \tag{2.3}$$

where

$$a = r\cos\theta,\ b = r\sin\theta,\ r = \sqrt{a^2+b^2},\ \theta = \tan^{-1}\frac{b}{a}. \tag{2.4}$$

e in (2.3) is a constant, which is called Euler's number or Napier's constant. It is

$$\mathrm{e} = 2.71828\ldots. \tag{2.5}$$

The *complex conjugate* (共役複素数) of z is $\bar{z} = a - jb$. The difference between z and $\bar{z}$ is the sign of the imaginary part. More specifically[†2], z and $\bar{z}$ are symmetric with respect to the real axis.

Let a_1, a_2, b_1, b_2, r_1, r_2, θ_1, and θ_2 be real numbers, and z_1 $(= a_1 + jb_1 = r_1\mathrm{e}^{j\theta_1})$ and z_2 $(= a_2 + jb_2 = r_2\mathrm{e}^{j\theta_2})$ be two complex numbers. The *four arithmetic operations* (四則演算) for z_1 and z_2 are defined to be

- *Addition* (足し算):

$$z_1 + z_2 = (a_1 + jb_1) + (a_2 + jb_2) = (a_1 + a_2) + j(b_1 + b_2); \tag{2.6}$$

---

†1 Note that i is used as the imaginary unit in high-school mathematics. However, since i is used for current in electrical engineering, j is usually used in engineering to avoid confusion.

†2 more specifically：具体的に。

- *Subtraction* (引き算):

$$z_1 - z_2 = (a_1 + jb_1) - (a_2 + jb_2) = (a_1 - a_2) + j(b_1 - b_2); \tag{2.7}$$

- *Multiplication* (掛け算):

$$z_1 \times z_2 = (r_1 e^{j\theta_1}) \times (r_2 e^{j\theta_2}) = (r_1 r_2) e^{j(\theta_1 + \theta_2)}; \tag{2.8}$$

- *Division* (割り算) ($a_2 + jb_2 \neq 0$ and $r_2 \neq 0$):

$$z_1 \div z_2 = (r_1 e^{j\theta_1}) \div (r_2 e^{j\theta_2}) = \frac{r_1}{r_2} e^{j(\theta_1 - \theta_2)}. \tag{2.9}$$

de Moivre's formula (ド・モアブルの定理)

$$(\cos\theta + j\sin\theta)^n = \cos n\theta + j\sin n\theta, \text{ or } \left(e^{j\theta}\right)^n = e^{jn\theta} \tag{2.10}$$

holds for a real number θ and an integer n.

【Example 2.1】 Calculate the polar form of $z = 5\sqrt{3} + j5$ and the orthogonal form of $z = 2e^{-\pi/4}$.

Since $r = |z| = \sqrt{(5\sqrt{3})^2 + 5^2} = 10$ and $\theta = \tan^{-1}\dfrac{5}{5\sqrt{3}} = \tan^{-1}\dfrac{1}{\sqrt{3}} = \dfrac{\pi}{6}$, $z = 10e^{j\pi/6}$.

Since $x = 2\cos\left(-\dfrac{\pi}{4}\right) = \sqrt{2}$ and $y = 2\sin\left(-\dfrac{\pi}{4}\right) = -\sqrt{2}$, $z = \sqrt{2} - j\sqrt{2}$.

Let $z_0 = x_0 + jy_0$, $z_1 = x_1 + jy_1$, and a rotational angle be α. The complex-number forms of coordinate translations are

$$\text{Parallel translation: } z = z_1 + z_0, \tag{2.11}$$

$$\text{Rotational translation: } z_\alpha = z_0 e^{j\alpha}. \tag{2.12}$$

And

$$\text{Parallel translation}\quad \begin{cases} x = x_1 + x_0, \\ y = y_1 + y_0; \end{cases} \tag{2.13}$$

$$\text{Rotational translation}\quad \begin{cases} x_\alpha = x_0 \cos\alpha - y_0 \sin\alpha, \\ y_\alpha = x_0 \sin\alpha + y_0 \cos\alpha. \end{cases} \tag{2.14}$$

2.2 Differential and Integral Calculus

Calculus (微積分) has two branches: *differential calculus* (微分) and *integral calculus* (積分), which is widely used in engineering.

2.2.1 Differential Calculus

Differential calculus computes an instantaneous rate of change. For two points x and $(x + \Delta x)$ in **Figure 2.2**, the slope of a straight line between the points of $y = f(x)$, $(x, f(x))$ and $(x + \Delta x, f(x + \Delta x))$, is

$$\text{Slope} = \frac{\Delta y}{\Delta x} = \frac{f(x + \Delta x) - f(x)}{(x + \Delta x) - x} = \frac{f(x + \Delta x) - f(x)}{\Delta x}. \tag{2.15}$$

The derivative of $f(x)$ at x is defined by taking the limit as Δx tends to zero:

$$\dot{f}(x) = \frac{\mathrm{d}f(x)}{\mathrm{d}x} = \lim_{\Delta x \to 0} \frac{f(x + \Delta x) - f(x)}{\Delta x}. \tag{2.16}$$

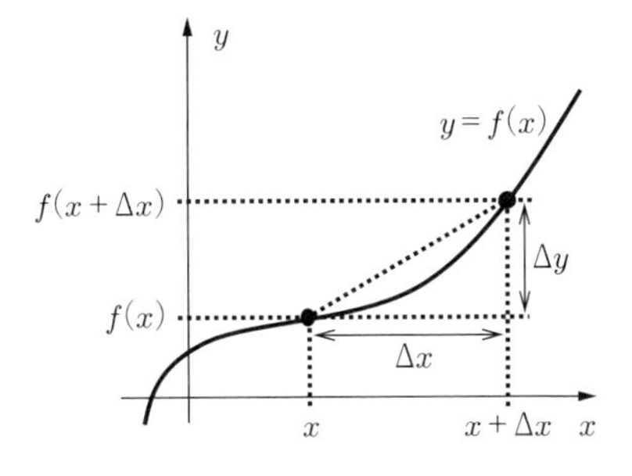

Figure 2.2 Slope of two points

【Example 2.2】 Calculate the derivative of $f(x) = x^2$ at $x = 3$.

According to the definition (2.16), we have

$$\begin{aligned} \dot{f}(x)|_{x=3} &= \lim_{\Delta x \to 0} \frac{(3+\Delta x)^2 - 3^2}{(3+\Delta x) - 3} = \lim_{\Delta x \to 0} \frac{\Delta x^2 + 6\Delta x}{\Delta x} \\ &= \lim_{\Delta x \to 0} (\Delta x + 6) = 6. \end{aligned} \tag{2.17}$$

The *derivative of a composite function* (合成関数の微分) $y = f(h(x))$ is

$$\frac{\mathrm{d}y}{\mathrm{d}x} = \frac{\mathrm{d}f(z)}{\mathrm{d}z} \cdot \frac{\mathrm{d}z}{\mathrm{d}x}, \ z = h(x). \tag{2.18}$$

Leibniz product rule (積の微分法則，ライプニッツ則) gives the derivative of the products of two functions:

$$\frac{\mathrm{d}}{\mathrm{d}x}[f(x)g(x)] = \frac{\mathrm{d}f(x)}{\mathrm{d}x}g(x) + f(x)\frac{\mathrm{d}g(x)}{\mathrm{d}x}. \tag{2.19}$$

The *derivatives of trigonometric functions* (三角関数の微分) are shown in **Table 2.1**.

Table 2.1 Derivatives of trigonometric functions

$f(x)$	$\mathrm{d}f(x)/\mathrm{d}x$
$\sin x$	$\cos x$
$\cos x$	$-\sin x$
$\tan x$	$\dfrac{1}{\cos^2 x}$
$\sec x \left(= \dfrac{1}{\cos x}\right)$	$\sec x \cdot \tan x$
$\csc x \left(= \dfrac{1}{\sin x}\right)$	$-\csc x \cdot \cot x$
$\cot x \left(= \dfrac{1}{\tan x}\right)$	$-\csc^2 x$

2.2.2 Indefinite and Definite Integral Calculus

Integral calculus is divided into *indefinite integrals* (不定積分) and *definite integrals* (定積分). The indefinite integrals represent the inverse operation of the differential calculus, and the definite integrals represent a method of calculating displacement, area, volume, and others.

For a given function $f(x)$, an indefinite integral of $f(x)$, $F(x)$, is a function that satisfies $\mathrm{d}F(x)/\mathrm{d}x = f(x)$. The indefinite integral of $f(x)$ is

$$\int f(x)\mathrm{d}x = F(x) + C, \tag{2.20}$$

where C is a constant. The definite integral of $f(x)$ is

$$\int_a^b f(x)\mathrm{d}x = F(b) - F(a). \tag{2.21}$$

It calculates the shaded area in **Figure 2.3**.

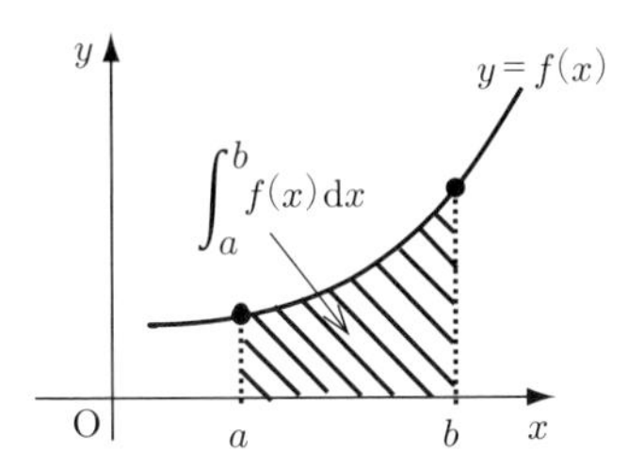

Figure 2.3 Concept of definite integral

Since $\mathrm{d}x^{n+1}/\mathrm{d}x = (n+1)x^n$, the indefinite integral of an exponential function is

$$\int x^n \mathrm{d}x = \frac{1}{n+1}x^{n+1} + C. \tag{2.22}$$

For example, if $f(x) = x^3$ then

$$\int x^3 \mathrm{d}x = \frac{1}{4}x^4 + C. \tag{2.23}$$

The *indefinite integrals of trigonometric functions* (三角関数の不定積分) are listed in **Table 2.2**, where C is a constant.

Integration by parts (部分積分法) is given by

$$\int_a^b f(x)\dot{g}(x)\mathrm{d}x = [f(x)g(x)]_a^b - \int_a^b \dot{f}(x)g(x)\mathrm{d}x. \tag{2.24}$$

Table 2.2 Integral formula of $f(x)$

$\int f(x)\mathrm{d}x$	$F(x)$
$\int \sin x \mathrm{d}x$	$-\cos x + C$
$\int \cos x \mathrm{d}x$	$\sin x + C$
$\int \sec^2 x \mathrm{d}x$	$\tan x + C$
$\int \csc^2 x \mathrm{d}x$	$-\cot x + C$
$\int \sec x \tan x \mathrm{d}x$	$\sec x + C$
$\int \csc x \cot x \mathrm{d}x$	$-\csc x + C$

or

$$\int_a^b f(x)\mathrm{d}g(x) = [f(x)g(x)]_a^b - \int_a^b g(x)\mathrm{d}f(x), \tag{2.25}$$

2.3 Partial Derivatives

Mapping a point (x, y) on the xy plane to a real number, z, yields a function of two variables: $z = f(x, y)$, where x and y are independent variables, and z is a dependent variable. For example, consider $z = f(x, y) = x^2 + 2xy + y^2 + 3$. The independent variables (x, y) are defined over the entire xy plane. In $z = f(x, y) = \sqrt{x^2 + y^2 - r^2}$, where r is a positive constant, the domain of the independent variable (x, y) is the entire area around and outside the circle $x^2 + y^2 = r^2$.

The function $f(x, y)$ of two variables becomes a function of only the variable x when the variable y is fixed. Also, when x is fixed, it becomes a function of y. It is possible to obtain the derivative of one independent variable by fixing the other. This is called the *partial derivative* (偏導関

数, 偏微分). The partial derivatives of $f(x,y)$ with respect to[†1] x and y are defined to be

$$\begin{cases} \dfrac{\partial f(x,y)}{\partial x} = \lim\limits_{\Delta x \to 0} \dfrac{f(x+\Delta x, y) - f(x,y)}{\Delta x}, \\ \dfrac{\partial f(x,y)}{\partial y} = \lim\limits_{\Delta y \to 0} \dfrac{f(x, y+\Delta y) - f(x,y)}{\Delta y}. \end{cases} \tag{2.26}$$

When there exist partial derivatives of a function $f(x,y)$ with respect to x and y and both are continuous, there exists a *total derivative* (全微分) $\mathrm{d}f$:

$$\mathrm{d}f = \frac{\partial f(x,y)}{\partial x}\mathrm{d}x + \frac{\partial f(x,y)}{\partial y}\mathrm{d}y. \tag{2.27}$$

The necessary and sufficient condition for $M(x,y)\mathrm{d}x + N(x,y)\mathrm{d}y$ to become a total derivative of a function $f(x,y)$ is that the following equation holds:

$$\frac{\partial M(x,y)}{\partial y} = \frac{\partial N(x,y)}{\partial x}. \tag{2.28}$$

This total derivative $M(x,y)\mathrm{d}x + N(x,y)\mathrm{d}y$ is called an *exact differential* (完全微分).

2.4 Exponential and Logarithmic Functions

The *exponential and logarithmic functions* (指数関数と対数関数) are symmetric with respect to the straight line $y = x$ (**Figure 2.4**). In other words, $y = a^x$ and $y = \log_a x$ are inverse functions to each other.

The derivatives of exponential and logarithmic functions are listed in **Table 2.3**.

The *indefinite integrals of exponential and logarithmic functions* (指数関数と対数関数の不定積分) are listed in **Table 2.4**[†2].

---

†1 with respect to ～：～に関して，～について（は）。

†2 The natural logarithm $\log_{\mathrm{e}} x$ is usually indicated by $\ln x$ or $\log x$.

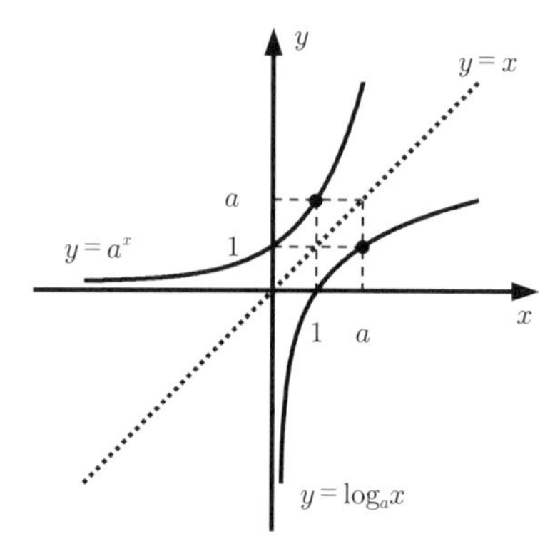

Figure 2.4 Exponential and logarithmic functions ($a > 1$)

Table 2.3 Derivatives of exponential and logarithmic functions

$f(x)$	$\mathrm{d}f(x)/\mathrm{d}x$
e^x	e^x
a^x	$a^x \ln a$
$\mathrm{e}^{f(x)}$	$\mathrm{e}^{f(x)} \dfrac{\mathrm{d}f(x)}{\mathrm{d}x}$
$a^{f(x)}$	$\ln a \cdot a^{f(x)} \dfrac{\mathrm{d}f(x)}{\mathrm{d}x}$
$\ln x$	$\dfrac{1}{x}$
$\log_a x$	$\dfrac{1}{\ln a} \dfrac{1}{x}$
$\ln \lvert f(x) \rvert$	$\dfrac{1}{f(x)} \dfrac{\mathrm{d}f(x)}{\mathrm{d}x}$

The Gaussian function $f(x) = \mathrm{e}^{-x^2}$ is widely used in many fields, for example, to describe a normal distribution in statistics and to define a Gaussian filter in signal processing. The definite integral of the Gaussian integral (also called the Euler-Poisson integral) is

$$\int_{-\infty}^{\infty} \mathrm{e}^{-x^2} \,\mathrm{d}x = \sqrt{\pi}, \quad \int_{0}^{\infty} \mathrm{e}^{-x^2} \,\mathrm{d}x = \frac{\sqrt{\pi}}{2}. \tag{2.29}$$

And the definite integral of a parametric Gaussian function is

Table 2.4 Indefinite integrals of exponential and logarithmic functions

$\int f(x)\mathrm{d}x$	$F(x)$
$\int \mathrm{e}^{kx}\mathrm{d}x$	$\frac{1}{k}\mathrm{e}^{kx} + C$
$\int a^x \mathrm{d}x$	$\frac{1}{\ln a}a^x + C$
$\int \ln x \mathrm{d}x$	$x \ln x - x + C$

$$\int_{-\infty}^{\infty} \mathrm{e}^{-a(x+b)^2}\, \mathrm{d}x = \sqrt{\frac{\pi}{a}}. \tag{2.30}$$

2.5 Trigonometric Functions

The *SI unit* (国際単位系, SI) of an angle is radian, which is defined to be the angle that subtends the arc of length equal to the radius of the circle (**Figure 2.5**).

$$180^\circ = \pi \text{ rad}, \quad \pi = 3.14159. \tag{2.31}$$

We use a unit circle (**Figure 2.6**) to define trigonometric functions as

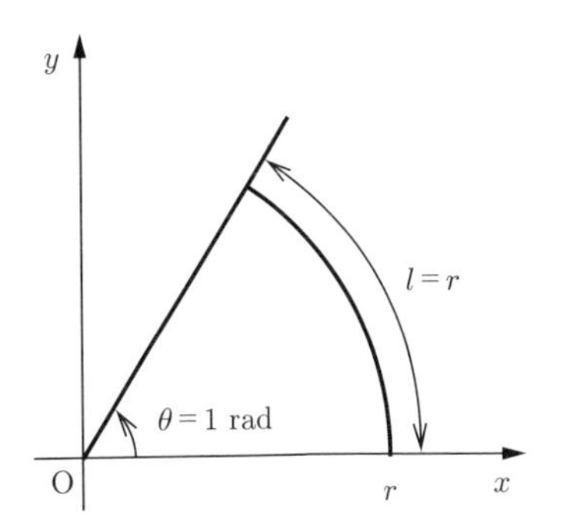

Figure 2.5 Definition of one radian

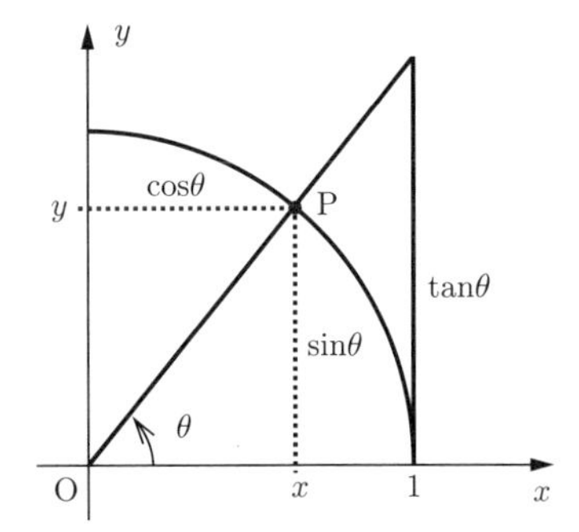

Figure 2.6 Definition of trigonometric functions

$$\cos\theta = \frac{x}{1} = x, \quad \sin\theta = \frac{y}{1} = y, \quad \tan\theta = \frac{\sin\theta}{\cos\theta} = \frac{y}{x}. \tag{2.32}$$

Letting a and b be non-zero real constants and using a and b to construct a right-angled triangle (**Figure 2.7**) yield

$$\sin\phi = \frac{b}{\sqrt{a^2+b^2}}, \quad \cos\phi = \frac{a}{\sqrt{a^2+b^2}}. \tag{2.33}$$

Thus, we can write $a\sin\theta + b\cos\theta$ as

$$\begin{aligned} a\sin\theta + b\cos\theta &= \sqrt{a^2+b^2}\left(\frac{a}{\sqrt{a^2+b^2}}\sin\theta + \frac{b}{\sqrt{a^2+b^2}}\cos\theta\right) \\ &= \sqrt{a^2+b^2}\,(\cos\phi\sin\theta + \sin\phi\cos\theta) \\ &= \sqrt{a^2+b^2}\,\sin(\theta+\phi), \end{aligned} \tag{2.34}$$

$$\phi = \tan^{-1}\frac{b}{a}. \tag{2.35}$$

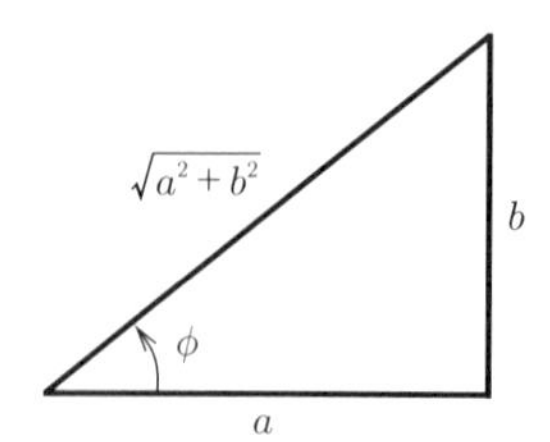

Figure 2.7 Construction of right-angled triangle

2.6 Various Functions

This section explains three types of functions: Hyperbolic functions, a Heaviside step function (step function), and a unit impulse function.

2.6.1 Hyperbolic Functions

Hyperbolic functions are analogs of† trigonometric functions. It is known

† analog of ～：～の類似物（相似形）。

that a suspension line (or a catenary curve) is described by a hyperbolic function $\cosh x$. A sigmoid function, which is widely used in many engineering fields, is also described by a hyperbolic function, $\tanh x$.

The following three functions are the *hyperbolic functions* (双曲線関数):

$$\begin{cases} \sinh ax = \dfrac{\mathrm{e}^{ax} - \mathrm{e}^{-ax}}{2}, \\ \cosh ax = \dfrac{\mathrm{e}^{ax} + \mathrm{e}^{-ax}}{2}, \\ \tanh ax = \dfrac{\mathrm{e}^{ax} - \mathrm{e}^{-ax}}{\mathrm{e}^{ax} + \mathrm{e}^{-ax}}, \end{cases} \tag{2.36}$$

where a is a constant. **Figure 2.8** shows these functions for $a = 1$.

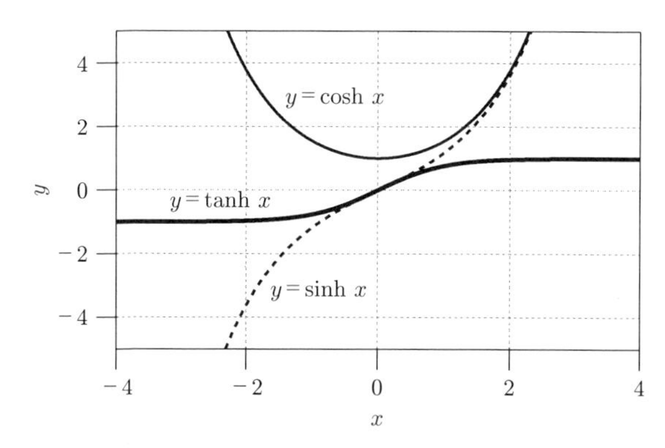

Figure 2.8 Hyperbolic functions

The definitions and basic characteristics are listed in **Table 2.5**. It also shows the similarities and differences between trigonometric and hyperbolic functions.

The differential and indefinate integral calculus of trigonometric and hyperbolic functions are listed in **Table 2.6**, where C is a constant.

Table 2.5 Definitions and basic characteristics of trigonometric and hyperbolic functions

Trigonometric function	Hyperbolic function
$e^{jx} = \cos x + j \sin x$	$e^{x} = \cosh x + \sinh x$
$\sin x = \dfrac{e^{jx} - e^{-jx}}{2j}$	$\sinh x = \dfrac{e^{x} - e^{-x}}{2}$
$\cos x = \dfrac{e^{jx} + e^{-jx}}{2}$	$\cosh x = \dfrac{e^{x} + e^{-x}}{2}$
$\tan x = \dfrac{\sin x}{\cos x}$	$\tanh x = \dfrac{\sinh x}{\cosh x}$
$\sin(-x) = -\sin x$	$\sinh(-x) = -\sinh x$
$\cos(-x) = \cos x$	$\cosh(-x) = \cosh x$
$\tan(-x) = -\tan x$	$\tanh(-x) = -\tanh x$
$\cos^2 x + \sin^2 x = 1$	$\cosh^2 x - \sinh^2 x = 1$
$\sin(x \pm y)$	$\sinh(x \pm y)$
$= \sin x \cos x \pm \cos x \sin y$	$= \sinh x \cosh x \pm \cosh x \sinh y$
$\cos(x \pm y)$	$\cosh(x \pm y)$
$= \cosh x \cos y \mp \sin x \sin y$	$= \cosh x \cosh y \pm \sinh x \sinh y$
$\tan(x \pm y) = \dfrac{\tan x \pm \tan y}{1 \mp \tan x \tan y}$	$\tanh(x \pm y) = \dfrac{\tanh x \pm \tanh y}{1 \pm \tanh x \tanh y}$

Table 2.6 Differential and indefinate integral calculus of trigonometric and hyperbolic functions

Trigonometric function	Hyperbolic function
$\sin x \leqq x \leqq \tan x$	$\sinh x \geqq x \geqq \tanh x$
$\lim\limits_{x \to 0} \dfrac{\sin x}{x} = 1$	$\lim\limits_{x \to 0} \dfrac{\sinh x}{x} = 1$
$\mathrm{d} \sin x / \mathrm{d}x = \cos x$	$\mathrm{d} \sinh x / \mathrm{d}x = \cosh x$
$\mathrm{d} \cos x / \mathrm{d}x = -\sin x$	$\mathrm{d} \cosh x / \mathrm{d}x = \sinh x$
$\mathrm{d} \tan x / \mathrm{d}x = \dfrac{1}{\cos^2 x}$	$\mathrm{d} \tanh x / \mathrm{d}x = \dfrac{1}{\cosh^2 x}$
$\displaystyle\int \sin x \mathrm{d}x = -\cos x + C$	$\displaystyle\int \sinh x \mathrm{d}x = \cosh x + C$
$\displaystyle\int \cos x \mathrm{d}x = \sin x + C$	$\displaystyle\int \cosh x \mathrm{d}x = \sinh x + C$
$\displaystyle\int \tan x \mathrm{d}x = -\ln \cos x + C$	$\displaystyle\int \tanh x \mathrm{d}x = \ln(\cosh x) + C$

2.6.2 Heaviside Step Function

The *Heaviside step function* (ヘヴィサイドのステップ関数) or the *unit step function* (単位ステップ信号), $1(t)$, is defined to be

$$1(t) = \begin{cases} 0, & \text{if } t < 0, \\ 1, & \text{if } t \geqq 0. \end{cases} \tag{2.37}$$

If the rising edge of the unit step signal is at t_0, then the Heaviside step function is

$$1(t - t_0) = \begin{cases} 0, & \text{if } t < t_0, \\ 1, & \text{if } t \geqq t_0. \end{cases} \tag{2.38}$$

It follows from (2.38) that a square wave can be described by

$$1_s(t) = 1(t - t_1) - 1(t - t_2) = \begin{cases} 0, & \text{if } t < t_1 \text{ or } t > t_2, \\ 1, & \text{if } t_1 \leqq t \leqq t_2. \end{cases} \tag{2.39}$$

Thus, multiplying an arbitrarily function, $f(t)$, by the above equation (2.39) extracts $f(t)$ for the period $[t_1, t_2]$:

$$f(t)1_s(t) = \begin{cases} 0, & \text{if } t < t_1 \text{ or } t > t_2, \\ f(t), & \text{if } t_1 \leqq t \leqq t_2. \end{cases} \tag{2.40}$$

2.6.3 Dirac Delta Function

The *Dirac delta function* (ディラックのデルタ関数) or the *unit impulse function* (単位インパルス関数), $\delta(t)$, is defined to be

$$\delta(t) = \begin{cases} \infty, & \text{if } t = 0, \\ 0, & \text{if } t \neq 0 \end{cases} \quad \text{and} \quad \int_{-\infty}^{\infty} \delta(t) \mathrm{d}t = 1. \tag{2.41}$$

Let

$$\delta_\epsilon(t) = \begin{cases} \dfrac{1}{2\epsilon}, & \text{if } -\epsilon < t \leqq \epsilon, \\ 0, & \text{otherwise.} \end{cases} \tag{2.42}$$

$\delta(t)$ is defined to be $\delta(t) = \lim_{\epsilon \to 0} \delta_\epsilon(t)$ (**Figure 2.9**). It is used to describe a shock.

Let a $(\neq 0)$ be a constant and $\phi(t)$ be a testing function satisfying $\lim_{t \to \pm\infty} \phi(t) = 0$. The following holds:

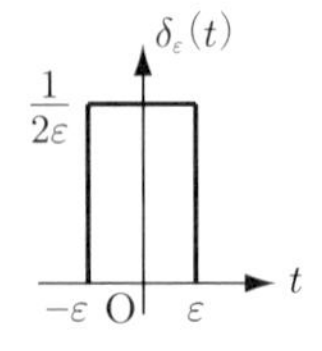

Figure 2.9 Definition of $\delta_\epsilon(t)$

(a) $$\int_{-\infty}^{\infty} \phi(t)1(t)\mathrm{d}t = \int_{0}^{\infty} \phi(t)\mathrm{d}t. \tag{2.43}$$

(b) $$\delta(-t) = \delta(t). \tag{2.44}$$

(c) $$\int_{-\infty}^{\infty} \delta(at)\phi(t)\mathrm{d}t = \frac{1}{|a|}\phi(0). \tag{2.45}$$

(d) $$\int_{-\infty}^{\infty} \phi(t)\delta(t-t_0)\mathrm{d}t = \phi(t_0). \tag{2.46}$$

(e) $$\delta(at) = \frac{1}{|a|}\delta(t). \tag{2.47}$$

(f) $$\delta(-t) = \delta(t). \tag{2.48}$$

(g) $$t\delta(t) = 0. \tag{2.49}$$

Problems

Basic Level

【1】 Calculate the Cartesian coordinates of the orthogonal form, the polar coordinates, and the conjugate numbers of the following complex numbers:

A. $\sqrt{2} + j\sqrt{2}$. B. $\dfrac{1}{2+j2}$.

C. $\dfrac{1}{j} - \dfrac{3j}{1-j}$. D. $j^3 - 4j^{21} + j$.

【2】 Calculate

A. $a^3 + b^3 - 3ab$ for $a = \dfrac{1+j\sqrt{3}}{2}$ and $b = \dfrac{1-j\sqrt{3}}{2}$.

B. $x^4 - 3x^2 + 6x - 4$ for $x = \dfrac{1+j\sqrt{3}}{2}$.

[3] Find x and y such that $\dfrac{x+1+j(y-3)}{5+j3} = 1+j$ holds.

[4] Calculate the orthogonal and polar formulas for the following complex numbers:

A. -1.

B. $1-\cos\phi + j\sin\phi \ (0 \leqq \phi \leqq \pi)$.

C. $\dfrac{j2}{-1+j}$.

D. $\dfrac{(\cos 5\phi + j\sin 5\phi)^2}{(\cos 3\phi - j\sin 3\phi)^3}$.

[5] Calculate the following limits:

A. $\lim\limits_{x\to\infty} \dfrac{2x+3}{4x+1}$.

B. $\lim\limits_{x\to\infty} \dfrac{x^{2n}}{e^x}$.

C. $\lim\limits_{x\to 0} \dfrac{e^x-1}{x}$.

D. $\lim\limits_{x\to 0} \dfrac{\sin x}{x}$.

[6] Differentiate the following functions:

A. $\sin x$

B. $\cos x$.

C. $\sin x + \ln x$.

D. $\sin 2x$.

E. $x\sin x$.

F. $x^n e^x$.

[7] Find the partial derivatives of the following functions:

A. $\dfrac{\partial(x^2y+y^4)}{\partial x}$.

B. $\dfrac{\partial(xy^2+y^4\cos x)}{\partial x}$.

C. $\dfrac{\partial e^{-(x^2+y^2)}}{\partial x}$.

[8] Calculate the indefinate integrals of

A. $\displaystyle\int \sin x dx$.

B. $\displaystyle\int e^x \sin x dx$.

C. $\displaystyle\int \log_{10} x dx$.

D. $\displaystyle\int e^{x+2} dx$.

[9] Find α and β for a function $f(t) = \alpha t^n + \beta$ that satisfies $f(2) = 17$, $f(4) = 77$, and $f(8) = 377$.

[10] Prove $0.3 < \log_{10} 2 < 0.4$.

[11] Convert the following angles from degrees to radians, and vice versa:

A. 30°.

B. 150°.

C. 225°.

D. $2\,\text{rad}$.

E. $\pi\,\text{rad}$.

F. $0.5\,\text{rad}$.

[12] Calculate the values of the following trigonometric functions:

A. $\sin 120^\circ$.

B. $\tan 210^\circ$.

C. $\sin 75^\circ - \sin 15^\circ$.

D. $\tan 210^\circ - \cos 330^\circ$.

[13] Find the period of $f(t) = \cos\dfrac{t}{2} + \cos\dfrac{t}{7}$.

[14] Prove (2.43)–(2.48).

Advanced Level

【1】 Find the minimum and maximum of $f(\theta) = \dfrac{2 - \sin\theta}{2 - \cos\theta}$ when $0 \leqq \theta < 2\pi$.

【2】 Prove

A. If $m = a^x$, $n = a^y$, and $m^y n^x = a^{2/z}$, then $xyz = 1$ for $a \neq 1$.

B. If $8^x = 9^y = 6^z$, then $\dfrac{2}{x} + \dfrac{3}{y} = \dfrac{6}{z}$.

【3】 When x and y satisfy $\log_x y = 2 + 3\log_y x$,

A. calculate $\log_x y$ and

B. describe the relationship as $y = f(x)$.

【4】 Find the maximum constant of a that satisfies $x^{\log_2 x} \geqq ax^2$ for $x > 0$.

【5】 Prove (2.49).

3 Fourier Series

There are many periodic functions in our daily lives, for example, the movement of clock hands and voltage supply by a commercial power source. This chapter explains how to use a Fourier series to describe a periodic function and some important properties of this expression.

3.1 Periodic Phenomena

A function, $y(t)$, is called a periodic function if it satisfies

$$y(t+T) = y(t) \tag{3.1}$$

for all t, where T is called the *period* (周期) of the function (**Figure 3.1**). The *frequency* (周波数) and *angular frequency* (角周波数) of the signal are

$$f = \frac{1}{T} \text{ [Hz]}, \ \omega = 2\pi f = \frac{2\pi}{T} \text{ [rad/s]}, \tag{3.2}$$

respectively.

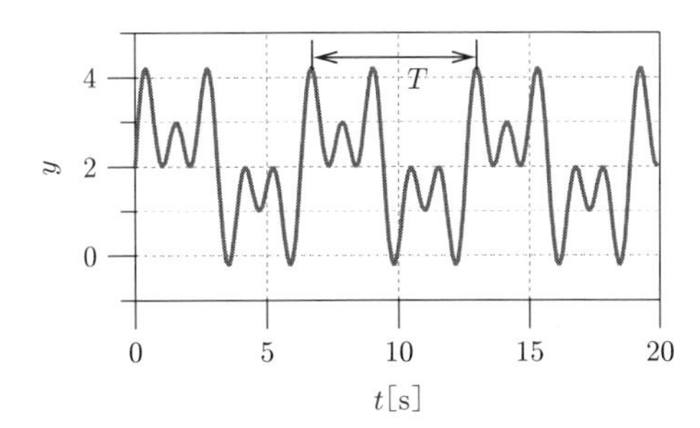

Figure 3.1 An example of periodic functions

Among periodic functions, $\sin\omega t$ and $\cos\omega t$ are two well-known ones. Consider a vector with its length of r on a complex plane and defined x and y axes as the horizontal and vertical projections (**Figure 3.2**). Set the initial position of the vector to $x = r$ and $y = 0$. If we turn the vector around the origin in the *counterclockwise direction* (反時計方向) with a rotational speed (or an angular frequency) of ω, then the position of the vector is

$$x(t) = r\cos\omega t, \quad y(t) = r\sin\omega t. \tag{3.3}$$

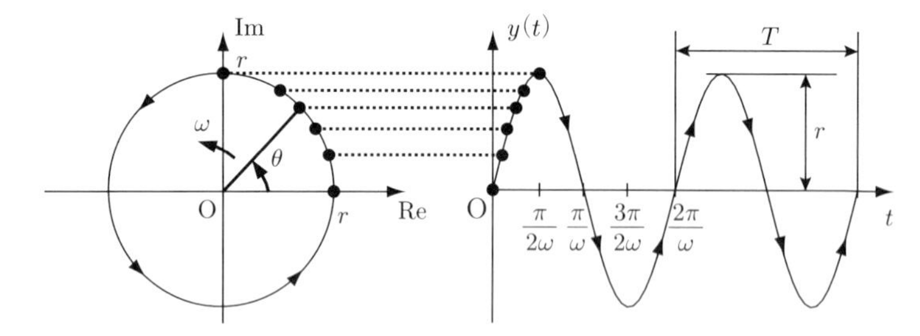

Figure 3.2 Explanation of a sine wave

Geometrically speaking, a sine function is symmetric about the origin, that is, its graph remains unchanged when it is rotated 180 degrees around the origin (it is called point symmetry). This relationship is described as

$$f(-t) = -f(t). \tag{3.4}$$

Such a function is called an *odd function* (奇関数). On the other hand, a cosine function is symmetric about the vertical axis, that is, its graph remains unchanged after reflection on the vertical axis (it is called axial symmetry). This relationship is described as

$$f(-t) = f(t). \tag{3.5}$$

Such a function is called an *even function* (偶関数). These are shown in **Figure 3.3**.

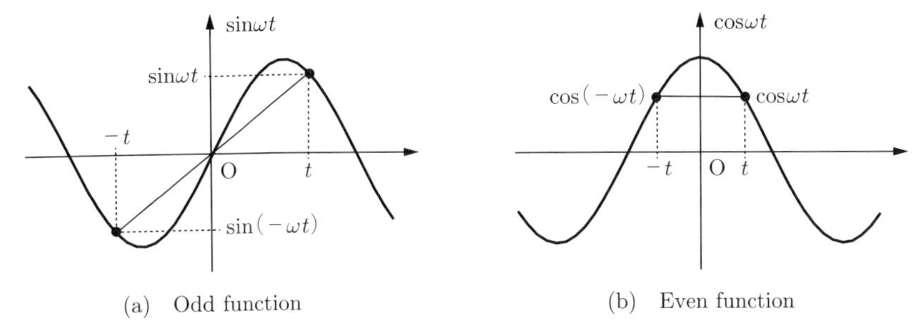

Figure 3.3 Odd and even functions

The basic properties of odd and even functions are as follows:

(1) The sum (or difference) of two odd functions is odd and the sum (or difference) of two even functions is even:

- Odd function $\pm$ odd function $\Longrightarrow$ odd function
- Even function $\pm$ even function $\Longrightarrow$ even function

(2) The product (or quotient) of two odd functions is even, the product (or quotient) of two even functions is even, and the product (or quotient) of an even and an odd functions is odd:

- Odd function $\times/$ odd function $\Longrightarrow$ even function
- Even function $\times/$ even function $\Longrightarrow$ even function
- Even function $\times/$ odd function $\Longrightarrow$ odd function

Given a function $f(t)$, if we define

$$f_e(t) = \frac{f(t) + f(-t)}{2}, \quad f_o(t) = \frac{f(t) - f(-t)}{2}, \tag{3.6}$$

then it is easy to verify that $f_e(t)$ is an even function and $f_o(t)$ is an odd function. Moreover, since

$$f(t) = f_e(t) + f_o(t), \tag{3.7}$$

any function can be decomposed as the sum of an even and an odd functions.

Sine and cosine functions play important roles in Fourier analysis, the following relationships of trigonometrical functions are of special importance in this book.

$$\sin(\alpha \pm \beta) = \sin\alpha\cos\beta \pm \cos\alpha\sin\beta, \tag{3.8}$$

$$\cos(\alpha \pm \beta) = \cos\alpha\cos\beta \mp \sin\alpha\sin\beta, \tag{3.9}$$

$$2\sin\alpha\cos\beta = \sin(\alpha+\beta) + \sin(\alpha-\beta), \tag{3.10}$$

$$2\sin\alpha\sin\beta = \cos(\alpha-\beta) - \cos(\alpha+\beta), \tag{3.11}$$

$$2\cos\alpha\cos\beta = \cos(\alpha+\beta) + \cos(\alpha-\beta), \tag{3.12}$$

$$\int \sin t dt = -\cos t + C, \quad \int \cos t dt = \sin t + C, \tag{3.13}$$

$$\int \cos t \sin t dt = -\frac{1}{2}\cos^2 t + C, \tag{3.14}$$

$$\int \sin^2 t dt = -\frac{1}{2}t - \frac{1}{4}\sin 2t + C, \tag{3.15}$$

$$\int \cos^2 t dt = \frac{1}{2}t + \frac{1}{4}\sin 2t + C, \tag{3.16}$$

$$\int t\sin nt \sin t dt = \frac{1}{n^2}\sin nt - \frac{1}{n}t\cos nt + C, \tag{3.17}$$

$$\int t\cos nt \sin t dt = \frac{1}{n^2}\cos nt + \frac{1}{n}t\sin nt + C, \tag{3.18}$$

where α, β, and t are variables, and C and n ($\neq 0$) are constants.

3.2 Expression of a Periodic Function

In physics and chemistry, we break down an analysis target into elements and examine it. In the same manner†, we decompose a periodic function into the sum of trigonometric functions, which is called a *Fourier series*

† in the same manner：同じように，同様に。

(フーリエ級数), to analyze such a function.

We describe a periodic function $f(t)$ with its period of T as

$$f_T(t) = \frac{a_0}{2} + \sum_{n=1}^{\infty} (a_n \cos n\omega t + b_n \sin n\omega t), \tag{3.19}$$

where $\omega = 2\pi/T$ [rad/s] is the angular frequency of $f_T(t)$, and a_n ($n = 0, 1, \ldots, \infty$) and b_n ($n = 1, 2, \ldots, \infty$) are Fourier coefficients that are given by

$$\begin{cases} a_n = \dfrac{2}{T} \displaystyle\int_{-T/2}^{T/2} f_T(t) \cos n\omega t \mathrm{d}t, \ n = 0, 1, 2, \ldots, \\ b_n = \dfrac{2}{T} \displaystyle\int_{-T/2}^{T/2} f_T(t) \sin n\omega t \mathrm{d}t, \ n = 1, 2, \ldots. \end{cases} \tag{3.20}$$

The first term, $a_0/2$, is the *direct current* (*DC*) *component* (直流成分) of $f(t)$ (that is, $\omega = 0$). Note that a coefficient, $1/2$, was used for a_0 to ensure that the formula for a_n ($n = 1, 2, \ldots, \infty$) also holds for a_0.

If two vectors, $x = [x_1, x_2, \ldots, x_m]^{\mathrm{T}}$ and $y = [y_1, y_2, \ldots, y_m]^{\mathrm{T}}$, are orthogonal, it means that their inner product $x^{\mathrm{T}} y$ is zero, that is, the sum of the product of each of their entries is zero ($x_1 y_1 + x_2 y_2 + \cdots + x_m y_m = 0$). We take a trigonometrical function (for example, $\sin \omega t$) as a vector with infinity entries (for example, $\sin \omega t_0, \sin \omega t_1, \sin \omega t_2, \cdots$). Then, it is not hard to understand that the orthogonality of trigonometrical functions over one period—their inner product is zero—is that the integral of the product of two different trigonometrical functions over one period is zero.

Orthogonality is the most important, interesting, also beautiful characteristic of trigonometrical functions. Now, we state the concept of orthogonality as follows:

Orthogonal functions (直交関数): If two functions $\phi_n(t)$ and $\phi_m(t)$ in a set $\{\phi_k(t)\}$ satisfy

$$\int_a^b \phi_n(t)\phi_m(t)\mathrm{d}t = \begin{cases} 0, & \text{if } n \neq m, \\ r_n, & \text{if } n = m, \end{cases} \tag{3.21}$$

where $r_n \neq 0$, then we call the set $\{\phi_k(t)\}$ orthogonal on the interval $a \leqq t \leqq b$. Moreover, if $r_n = 1$ $(n = 1, 2, \ldots)$, then the set is called *orthonormal* (正規直交) on this interval.

It is easy to verify that, for any $n, m = 0, 1, 2, \ldots$ and $n \neq m$,

$$\begin{cases} \displaystyle\int_{-T/2}^{T/2} \cos n\omega t \cos m\omega t \mathrm{d}t = \int_0^T \cos n\omega t \cos m\omega t \mathrm{d}t = 0, \\ \displaystyle\int_{-T/2}^{T/2} \sin n\omega t \sin m\omega t \mathrm{d}t = \int_0^T \sin n\omega t \sin m\omega t \mathrm{d}t = 0, \\ \displaystyle\int_{-T/2}^{T/2} \sin n\omega t \cos m\omega t \mathrm{d}t = \int_0^T \sin n\omega t \cos m\omega t \mathrm{d}t = 0, \\ \displaystyle\int_{-T/2}^{T/2} \cos^2 n\omega t \mathrm{d}t = \int_0^T \cos^2 n\omega t \mathrm{d}t = \frac{T}{2}, \\ \displaystyle\int_{-T/2}^{T/2} \sin^2 n\omega t \mathrm{d}t = \int_0^T \sin^2 n\omega t \mathrm{d}t = \frac{T}{2} \end{cases} \tag{3.22}$$

hold. It is clear that the function group $\{1, \sin\omega t, \cos\omega t, \sin 2\omega t, \cos 2\omega t, \ldots\}$ is a set of orthogonal functions in the interval $-T/2 \leqq t \leqq T/2$ (or $0 \leqq t \leqq T$).

Now, we show the derivation of a_n $(n = 0, 1, 2, \ldots)$ in (3.20) based on the concept of orthogonality. First, we consider $n = 0$ and integrate each side of (3.19) yields

$$\begin{aligned} &\int_{-T/2}^{T/2} f_T(t)\mathrm{d}t \\ &= \int_{-T/2}^{T/2} \frac{a_0}{2}\mathrm{d}t + \sum_{n=1}^{\infty}\left(a_n \int_{-T/2}^{T/2} \cos n\omega t \mathrm{d}t + b_n \int_{-T/2}^{T/2} \sin n\omega t \mathrm{d}t\right). \end{aligned}$$

Since

$$\int_{-T/2}^{T/2} \cos n\omega t \mathrm{d}t = 0, \quad \int_{-T/2}^{T/2} \sin n\omega t \mathrm{d}t = 0, \quad n = 1, 2, \ldots, \tag{3.23}$$

we have

$$a_0 = \frac{2}{T}\int_{-T/2}^{T/2} f_T(t)\mathrm{d}t. \tag{3.24}$$

Then, we consider $n = 1, 2, \ldots$. Multiply both sides of (3.19) by $\cos m\omega t$ and integrating them gives

$$\int_{-T/2}^{T/2} f_T(t)\cos m\omega t\mathrm{d}t = \int_{-T/2}^{T/2} \frac{a_0}{2}\cos m\omega t\mathrm{d}t$$
$$+\sum_{n=1}^{\infty}\left(a_n\int_{-T/2}^{T/2}\cos n\omega t\cos m\omega t\mathrm{d}t + b_n\int_{-T/2}^{T/2}\sin n\omega t\cos m\omega t\mathrm{d}t\right).$$

It follows from the results given in (3.22) that

$$\int_{-T/2}^{T/2} f_T(t)\cos n\omega t\mathrm{d}t = a_n\int_{-T/2}^{T/2}\cos^2 n\omega t\mathrm{d}t = a_n\frac{T}{2}$$

when $m = n$. Thus,

$$a_n = \frac{2}{T}\int_{-T/2}^{T/2} f_T(t)\cos n\omega t\mathrm{d}t, \quad n = 1, 2, \ldots. \tag{3.25}$$

Combining (3.24) and (3.25) provides us with a_n $(n = 0, 1, 2, \ldots, \infty)$ in (3.20). The derivation of b_n $(n = 1, 2, \ldots, \infty)$ in (3.20) can be shown in the same manner.

【Example 3.1】 Expand a periodic function

$$f_T(t) = \begin{cases} 0, & \text{if } 0 < t < \dfrac{\pi}{2}, \\ 1, & \text{if } \dfrac{\pi}{2} < t < \pi. \end{cases} \tag{3.26}$$

in a Fourier series (**Figure 3.4**).

Since $T = \pi$ and $\omega = 2\pi/T = 2$,

$$a_0 = \frac{2}{T}\int_{-T/2}^{T/2} f_T(t)\mathrm{d}t = \frac{2}{\pi}\left(\int_{-\pi/2}^{0} 1\mathrm{d}t + \int_{0}^{\pi/2} 0\mathrm{d}t\right)$$
$$= \frac{2}{\pi}\int_{-\pi/2}^{0} 1\mathrm{d}t = 1, \tag{3.27}$$

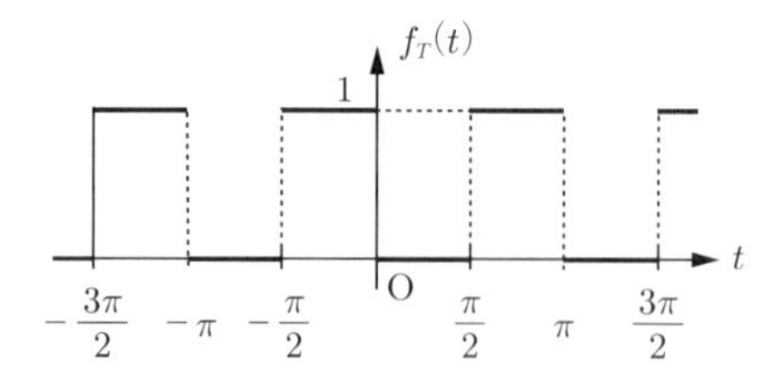

Figure 3.4 $f_T(t)$ in (3.26)

$$\begin{aligned} a_n &= \frac{2}{T}\int_{-T/2}^{T/2} f_T(t)\cos n\omega t dt \\ &= \frac{2}{\pi}\left(\int_{-\pi/2}^{0} 1\times\cos 2nt dt + \int_{0}^{\pi/2} 0\times\cos 2nt dt\right) \\ &= \frac{2}{\pi}\int_{-\pi/2}^{0}\cos 2nt dt = 0, n = 1, 2, \ldots, \end{aligned} \tag{3.28}$$

and

$$\begin{aligned} b_n &= \frac{2}{T}\int_{-T/2}^{T/2} f_T(t)\sin n\omega t dt \\ &= \frac{2}{\pi}\left(\int_{-\pi/2}^{0} 1\times\sin 2nt dt + \int_{0}^{\pi/2} 0\times\sin 2nt dt\right) \\ &= \frac{2}{\pi}\int_{-\pi/2}^{0}\sin 2nt dt = \left[\frac{-1}{n\pi}\cos 2nt\right]_{-\pi/2}^{0} = \frac{\cos n\pi - 1}{n\pi} \\ &= \frac{(-1)^n - 1}{n\pi}, n = 1, 2, \ldots. \end{aligned} \tag{3.29}$$

Hence,

$$f_T(t) = \frac{1}{2} - \frac{2}{\pi}\left(\sin 2t + \frac{1}{3}\sin 6t + \frac{1}{5}\sin 10t + \cdots\right). \tag{3.30}$$

The terms $\cos n\omega t$ and $\sin n\omega t$ are even and odd functions of t, respectively. An odd periodic function $f_{To}(t)$ and an even periodic function $f_{Te}(t)$ can be described as

$$f_{To}(t) = \sum_{n=1}^{\infty} b_n \sin n\omega t, \quad f_{Te}(t) = \frac{a_0}{2} + \sum_{n=1}^{\infty} a_n \cos n\omega t, \tag{3.31}$$

respectively, where a_0, a_n, and b_n are given by (3.20).

Note that the key point in the calculation of Fourier coefficients is to integrate the related functions for one period. Thus, they can also be calculated by

$$\begin{cases} a_n = \dfrac{2}{T}\displaystyle\int_0^T f(t)\cos n\omega t dt, \quad n = 0, 1, 2, \ldots, \\ b_n = \dfrac{2}{T}\displaystyle\int_0^T f(t)\sin n\omega t dt, \quad n = 1, 2, \ldots. \end{cases} \tag{3.32}$$

There is no problem to choose either $[-T/2, T/2]$ or $[0, T]$ for the integral interval.

If a periodic function is a sum of two other periodic functions

$$f_T(t) = f_T^{(1)}(t) + f_T^{(2)}(t) \tag{3.33}$$

and

$$\begin{cases} f_T^{(1)}(t) = \dfrac{a_0^{(1)}}{2} + \displaystyle\sum_{n=1}^{\infty}\left[a_n^{(1)}\cos n\omega t + b_n^{(1)}\sin n\omega t\right], \\ f_T^{(2)}(t) = \dfrac{a_0^{(2)}}{2} + \displaystyle\sum_{n=1}^{\infty}\left[a_n^{(2)}\cos n\omega t + b_n^{(2)}\sin n\omega t\right], \end{cases} \tag{3.34}$$

$$\begin{cases} a_n^{(1)} = \dfrac{2}{T}\displaystyle\int_{-T/2}^{T/2} f_T^{(1)}(t)\cos n\omega t dt, \ n = 0, 1, 2, \ldots, \\ b_n^{(1)} = \dfrac{2}{T}\displaystyle\int_{-T/2}^{T/2} f_T^{(1)}(t)\sin n\omega t dt, \ n = 1, 2, \ldots, \\ a_n^{(2)} = \dfrac{2}{T}\displaystyle\int_{-T/2}^{T/2} f_T^{(2)}(t)\cos n\omega t dt, \ n = 0, 1, 2, \ldots, \\ b_n^{(2)} = \dfrac{2}{T}\displaystyle\int_{-T/2}^{T/2} f_T^{(2)}(t)\sin n\omega t dt, \ n = 1, 2, \ldots, \end{cases} \tag{3.35}$$

then the Fourier coefficients of $f_T(t)$ are the sum of the corresponding Fourier coefficients of $f_T^{(1)}(t)$ and $f_T^{(2)}(t)$

$$\begin{cases} f_T(t) = \dfrac{a_0}{2} + \displaystyle\sum_{n=1}^{\infty} (a_n \cos n\omega t + b_n \sin n\omega t), \\ a_n = a_n^{(1)} + a_n^{(2)} \ (n = 0, 1, 2, \ldots), \ b_n = b_n^{(1)} + b_n^{(2)} \ (n = 1, 2, \ldots). \end{cases} \tag{3.36}$$

The Fourier coefficients of a periodic function $cf_T(t)$ (c is a constant) are c times the corresponding Fourier coefficients of $f_T(t)$.

【Example 3.2】 Find the Fouirer series of the function (**Figure 3.5**)

$$f_T(t) = t + \pi, \text{ if } -\pi \leqq t < \pi, \text{ and } f_T(t + 2\pi) = f_T(t). \tag{3.37}$$

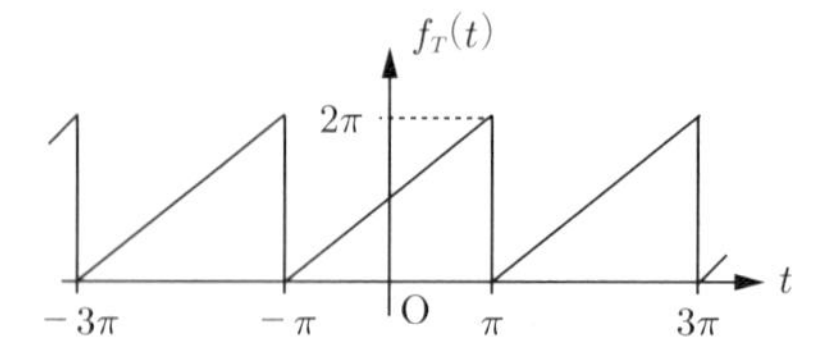

Figure 3.5 $f_T(t)$ in (3.37)

Note that $T = 2\pi$ and $\omega = 2\pi/T = 1$. We decompose the function as $f_T(t) = f_T^{(1)}(t) + f_T^{(2)}(t)$, where $f_T^{(1)}(t) = t$ and $f_T^{(2)}(t) = \pi$. The Fourier coefficients of $f_T^{(2)}(t)$ are zero except for $a_0^{(2)}$ $(= 2\pi)$. Since $f_T^{(1)}(t)$ is odd, $a_n^{(1)} = 0$ $(n = 0, 1, 2, \ldots)$ and

$$\begin{aligned} b_n^{(1)} &= \frac{2}{T} \int_{-T/2}^{T/2} f_T^{(1)}(t) \sin n\omega t dt = \frac{2}{2\pi} \int_{-\pi}^{\pi} t \sin nt dt \\ &= \frac{2}{\pi} \int_0^{\pi} t \sin nt dt = \frac{2}{\pi} \left(\left[\frac{-\pi \cos nt}{n} \right]_0^{\pi} + \frac{1}{n} \int_0^{\pi} \cos nt dt \right) \\ &= -\frac{2}{n} \cos n\pi = (-1)^{n+1} \frac{2}{n}. \end{aligned} \tag{3.38}$$

As a result, the Fourier coefficients of $f_T(t)$ are the sum of those of $f_T^{(1)}(t)$ and $f_T^{(2)}(t)$, that is,

$$\begin{cases} a_0 = a_0^{(1)} + a_0^{(2)} = 2\pi, \ a_n = a_n^{(1)} + a_n^{(2)} = 0, \\ b_n = b_n^{(1)} + b_n^{(2)} = (-1)^{n+1}\dfrac{2}{n}, \ n = 1, 2, \ldots. \end{cases} \tag{3.39}$$

Hence,

$$f_T(t) = \pi + 2\left(\sin t - \frac{1}{2}\sin 2t + \frac{1}{3}\sin 3t - \cdots\right). \tag{3.40}$$

There is the question of whether or not the Fourier series of $f_T(t)$ converges to the original function $f_T(t)$. The answer is always yes if the Fourier series of $f_T(t)$ satisfies some reasonably general hypotheses. More specifically, if we take the first $(2k+1)$ terms of (3.19) to be

$$f_{Tk}(t) = \frac{a_0}{2} + \sum_{n=1}^{k}\left(a_n \cos n\omega t + b_n \sin n\omega t\right), \tag{3.41}$$

we can prove that, if $f_T(t)$ satisfies the following two conditions on $[-T/2, T/2]$:

1) $f_T(t)$ is piecewise continuous (that is, it only has finitely many finite jump discontinuities) and

2) $f_T(t)$ is piecewise smooth [that is, $f_T(t)$ and its first derivative, $\mathrm{d}f_T(t)/\mathrm{d}t$, are both piecewise continuous],

then $f_{Tk}(t)$ converges to $f_T(t)$ as $k \to \infty$

If we use the first $(2k+1)$ terms of the Fourier series, (3.41), to approximate a periodic function, the approximation error is

$$e_k(t) = f_T(t) - f_{Tk}(t). \tag{3.42}$$

The *mean-square error* (平均二乗誤差) is defined to be

$$E_k = \frac{1}{T}\int_{-T/2}^{T/2} e_k^2(t)\mathrm{d}t = \frac{1}{T}\int_{-T/2}^{T/2} [f_T(t) - f_{Tk}(t)]^2 \mathrm{d}t. \tag{3.43}$$

The approximation (3.41) has the least mean-square error. To show this property, we substitute (3.41) into (3.43) and obtain

$$E_k = \frac{1}{T}\int_{-T/2}^{T/2}\left[f_T(t)-\frac{a_0}{2}-\sum_{n=1}^{k}(a_n\cos n\omega t+b_n\sin n\omega t)\right]^2 \mathrm{d}t. \tag{3.44}$$

To find an approximation (3.41) that has the least mean-square error, the partial derivatives of E_k with respect to a_0, a_n, and b_n have to satisfy

$$\frac{\partial E_k}{\partial a_0}=0,\quad \frac{\partial E_k}{\partial a_n}=0,\quad \frac{\partial E_k}{\partial b_n}=0,\quad n=1,2,\ldots,k. \tag{3.45}$$

Using the orthogonality (3.22) yields

$$\begin{aligned}\frac{\partial E_k}{\partial a_0} &= -\frac{1}{T}\int_{-T/2}^{T/2}\left[f_T(t)-\frac{a_0}{2}-\sum_{n=1}^{k}(a_n\cos n\omega t+b_n\sin n\omega t)\right]\mathrm{d}t\\ &= \frac{a_0}{2}-\frac{1}{T}\int_{-T/2}^{T/2}f_T(t)\mathrm{d}t=0,\end{aligned} \tag{3.46}$$

$$\begin{aligned}\frac{\partial E_k}{\partial a_n} &= -\frac{2}{T}\int_{-T/2}^{T/2}\left[f_T(t)-\frac{a_0}{2}-\sum_{n=1}^{k}(a_n\cos n\omega t+b_n\sin n\omega t)\right]\\ &\quad\times\cos n\omega t\mathrm{d}t\\ &= a_n-\frac{2}{T}\int_{-T/2}^{T/2}f_T(t)\cos n\omega t\mathrm{d}t=0,\end{aligned} \tag{3.47}$$

$$\begin{aligned}\frac{\partial E_k}{\partial b_n} &= -\frac{2}{T}\int_{-T/2}^{T/2}\left[f_T(t)-\frac{a_0}{2}-\sum_{n=1}^{k}(a_n\cos n\omega t+b_n\sin n\omega t)\right]\\ &\quad\times\sin n\omega t\mathrm{d}t\\ &= b_n-\frac{2}{T}\int_{-T/2}^{T/2}f_T(t)\sin n\omega t\mathrm{d}t=0.\end{aligned} \tag{3.48}$$

Clearly, (3.46), (3.47), and (3.48) show that the coefficients given in (3.20) ($n=0,1,\ldots,k$) have the least mean-square error for the approximation using the first $(2k+1)$ terms of the Fourier series.

3.3 Complex Form of Fourier Series

In the previous sections, real numbers were used in a Fourier series. That is, for a periodic function, $f_T(t)$, of period T, its Fourier series is given by (3.19). This section explores the use of complex numbers for a Fourier series. A complex form not only makes the mathematical expression of a Fourier series elegant, but also makes it easy to manipulate. For this reason, the complex form is common in writing a series expansion.

Recalling Euler's formula

$$\mathrm{e}^{j\theta} = \cos\theta + j\sin\theta, \tag{3.49}$$

we have

$$\cos\theta = \frac{\mathrm{e}^{j\theta} + \mathrm{e}^{-j\theta}}{2}, \quad \sin\theta = \frac{\mathrm{e}^{j\theta} - \mathrm{e}^{-j\theta}}{2j}. \tag{3.50}$$

Substituting (3.50) into (3.19) yields†

$$\begin{aligned} f_T(t) &= \frac{a_0}{2} + \sum_{n=1}^{\infty}\left(a_n\frac{\mathrm{e}^{jn\omega t} + \mathrm{e}^{-jn\omega t}}{2} + b_n\frac{\mathrm{e}^{jn\omega t} - \mathrm{e}^{-jn\omega t}}{2j}\right) \\ &= \frac{a_0}{2} + \sum_{n=1}^{\infty}\left(\frac{a_n - jb_n}{2}\mathrm{e}^{jn\omega t} + \frac{a_n + jb_n}{2}\mathrm{e}^{-jn\omega t}\right). \end{aligned} \tag{3.51}$$

Letting

$$c_0 = \frac{a_0}{2}, \quad c_n = \frac{a_n - jb_n}{2}, \quad c_{-n} = \frac{a_n + jb_n}{2} \tag{3.52}$$

yields

$$\begin{aligned} f_T(t) &= c_0 + \sum_{n=1}^{\infty}\left(c_n\mathrm{e}^{jn\omega t} + c_{-n}\mathrm{e}^{-jn\omega t}\right) \\ &= c_0 + \sum_{n=1}^{\infty} c_n\mathrm{e}^{jn\omega t} + \sum_{n=-1}^{-\infty} c_n\mathrm{e}^{jn\omega t}. \end{aligned} \tag{3.53}$$

That is,

$$f_T(t) = \sum_{n=-\infty}^{\infty} c_n\mathrm{e}^{jn\omega t}, \tag{3.54}$$

† Substituting A into B yields C：A を B に代入すると C を得る。

where

$$c_0 = \frac{a_0}{2} = \frac{2}{T}\int_{-T/2}^{T/2} f_T(t)\mathrm{d}t, \tag{3.55}$$

$$\begin{aligned} c_n &= \frac{a_n - jb_n}{2} \\ &= \frac{1}{T}\left[\int_{-T/2}^{T/2} f_T(t)\cos n\omega t\mathrm{d}t - j\int_{-T/2}^{T/2} f_T(t)\sin n\omega t\mathrm{d}t\right] \\ &= \frac{1}{T}\int_{-T/2}^{T/2} f_T(t)\left(\cos n\omega t - j\sin n\omega t\right)\mathrm{d}t \\ &= \frac{1}{T}\int_{-T/2}^{T/2} f_T(t)\mathrm{e}^{-jn\omega t}\mathrm{d}t, \quad n = 1, 2, \ldots, \infty, \end{aligned} \tag{3.56}$$

and

$$c_{-n} = \frac{a_n + jb_n}{2} = \frac{1}{T}\int_{-T/2}^{T/2} f_T(t)\mathrm{e}^{jn\omega t}\mathrm{d}t, \qquad n = -1, -2, \ldots, -\infty. \tag{3.57}$$

Let $f_T(t)$ be a real number. Then, $c_{-n} = c_n^*$ (conjugate complex number) holds.

Summarizing the above explanation, we have

$$c_n = \frac{1}{T}\int_{-T/2}^{T/2} f_T(t)\mathrm{e}^{-jn\omega t}\mathrm{d}t, \quad n = 0, \pm 1, \pm 2, \ldots. \tag{3.58}$$

We call (3.54) the complex Fourier series of the periodic function $f_T(t)$, and (3.58) the coefficients of the complex Fourier series. Note that

$$|c_n| = \frac{1}{2}\sqrt{a_n^2 + b_n^2}, \quad \phi_n = \tan^{-1}\left(-\frac{b_n}{a_n}\right). \tag{3.59}$$

On the other hand, for (3.19),

$$\begin{aligned} f_T(t) &= \frac{a_0}{2} + \sum_{n=1}^{\infty}\left(a_n\cos n\omega t + b_n\sin n\omega t\right) \\ &= \frac{a_0}{2} + \sum_{n=1}^{\infty}\sqrt{a_n^2 + b_n^2}\left(\frac{a_n}{\sqrt{a_n^2 + b_n^2}}\cos n\omega t + \frac{b_n}{\sqrt{a_n^2 + b_n^2}}\sin n\omega t\right) \end{aligned}$$

$$= \frac{a_0}{2} + \sum_{n=1}^{\infty} \sqrt{a_n^2 + b_n^2} \left(\frac{a_n}{\sqrt{a_n^2 + b_n^2}} \cos n\omega t - \frac{-b_n}{\sqrt{a_n^2 + b_n^2}} \sin n\omega t \right)$$

$$= \frac{a_0}{2} + \sum_{n=1}^{\infty} \sqrt{a_n^2 + b_n^2} \cos(n\omega t + \phi_n), \tag{3.60}$$

where

$$\sin \phi_n = \frac{-b_n}{\sqrt{a_n^2 + b_n^2}}, \quad \cos \phi_n = \frac{a_n}{\sqrt{a_n^2 + b_n^2}}, \quad \phi_n = \tan^{-1} \frac{-b_n}{a_n}. \tag{3.61}$$

So, we have

$$f_T(t) = \sum_{n=0}^{\infty} 2|c_n| \cos(n\omega t + \phi_n). \tag{3.62}$$

On the other hand, we can also write (3.60) as

$$\begin{aligned} f_T(t) &= \frac{a_0}{2} \\ &\quad + \sum_{n=1}^{\infty} \frac{\sqrt{a_n^2 + b_n^2}}{2} \left(\frac{a_n}{\sqrt{a_n^2 + b_n^2}} \cos n\omega t - \frac{-b_n}{\sqrt{a_n^2 + b_n^2}} \sin n\omega t \right) \\ &\quad + \sum_{-n=1}^{\infty} \frac{\sqrt{a_{-n}^2 + b_{-n}^2}}{2} \left[\frac{a_{-n}}{\sqrt{a_{-n}^2 + b_{-n}^2}} \cos(-n)\omega t \right. \\ &\quad \left. - \frac{-b_{-n}}{\sqrt{a_{-n}^2 + b_{-n}^2}} \sin(-n)\omega t \right] \\ &= \sum_{n=-\infty}^{\infty} \frac{\sqrt{a_n^2 + b_n^2}}{2} \cos(n\omega t + \phi_n). \end{aligned} \tag{3.63}$$

Thus,

$$f_T(t) = \sum_{n=-\infty}^{\infty} |c_n| \cos(n\omega t + \phi_n). \tag{3.64}$$

【Example 3.3】 Find the complex form of the Fourier series of the periodic function (**Figure 3.6**)

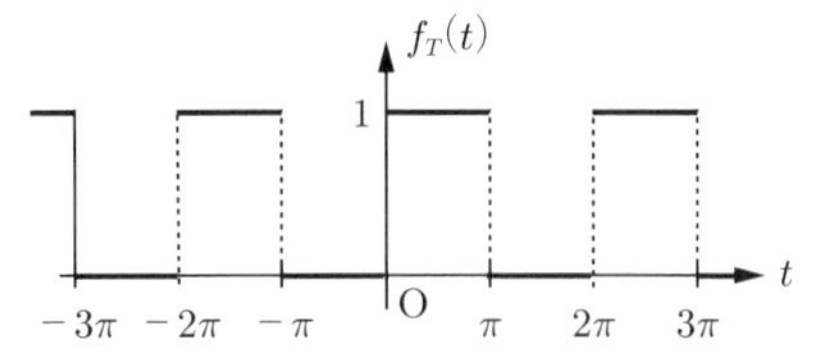

Figure 3.6 $f_T(t)$ in (3.65)

$$f_T(t) = \begin{cases} 0, & \text{if } -\pi < t < 0, \\ 1, & \text{if } 0 < t < \pi. \end{cases} \tag{3.65}$$

Since $T = 2\pi$ and $\omega = 2\pi/T = 1$,

$$c_0 = \frac{1}{T}\int_{-T/2}^{T/2} f_T(t)\mathrm{d}t = \frac{1}{2\pi}\int_{-\pi}^{\pi} f_T(t)\mathrm{d}t = \frac{1}{2\pi}\int_0^{\pi}\mathrm{d}t = \frac{1}{2},$$

$$\begin{aligned} c_n &= \frac{1}{T}\int_{-T/2}^{T/2} f_T(t)\mathrm{e}^{-jn\omega t}\mathrm{d}t = \frac{1}{2\pi}\int_{-\pi}^{\pi} f_T(t)\mathrm{e}^{-jnt}\mathrm{d}t \\ &= \frac{1}{2\pi}\int_0^{\pi}\mathrm{e}^{-jnt}\mathrm{d}t = \frac{1}{2\pi}\times\left.\frac{\mathrm{e}^{-jnt}}{-jn}\right|_0^{\pi} \\ &= \frac{1}{2\pi}\frac{\mathrm{e}^{-jn\pi}-1}{-jn} = -j\frac{1-(-1)^n}{2n\pi},\ n = \pm 1, \pm 2, \pm 3, \ldots, \end{aligned}$$

and

$$|c_0| = \frac{1}{2},\ \phi_0 = 0,\ |c_n| = \frac{1}{n\pi},\ \phi_n = -90^\circ,\ n = \pm 1, \pm 3, \pm 5, \ldots.$$

It is clear from (3.64) that[†] $|c_n|$ is the amplitude of the n-th harmonic wave of the periodic function, which is called the *amplitude spectrum* (振幅スペクトル), and ϕ_n is the phase shift of the n-th harmonic wave, which is called the *phase spectrum* (位相スペクトル). Since a spectrum distributes over the frequency range $(-\infty, \infty)$, it is called a *two-sided spectrum* (両側スペクトル). On the other hand, as shown in (3.62), most real-world frequency-analysis instruments just show the positive half of the frequency

† It is clear from A that ～：～ことは A から明らかである。

spectrum [$\omega \in [0, \infty)$] because the spectrum of a real-world signal is symmetrical around the DC component and the negative frequency information is redundant. This spectrum is called a *single-sided spectrum* (片側スペクトル), which is twice as large as that of a two-sided one.

Problems

Basic Level

【1】 Choose odd and even functions from the following list:

A. $\sin t$. B. e^t. C. $\cos t$. D. $5t^2$. E. $2t^2 + 3$. F. $\dfrac{8}{t^3}$.

G. $\dfrac{\sin t}{t}$. H. $at^n + b$ ($a \neq 0, n = 1, 2, \ldots, N,\ N < \infty$).

【2】 Is $f(t) = \cos t + \cos(1 + \pi)t$ periodic?

【3】 Find the periods of the following functions:

A. $\sin nt$. B. $\cos 4\pi t$. C. $\cos \dfrac{\pi t}{k}$. D. $|\cos t|$.

E. $5 \sin t + \sin \dfrac{t}{2} + \sin \dfrac{t}{4}$.

【4】 For the following functions, check whether they are continuous, piecewise continuous, or piecewise smooth on $[-\pi, \pi]$.

A. $f(t) = \dfrac{1}{\sin t}$. B. $f(t) = (\sin t)^{1/3}$. C. $f(t) = (\sin t)^{4/3}$.

D. $f(t) = \begin{cases} \cos t, & \text{if } t > 0, \\ -\cos t, & \text{if } t \leqq 0. \end{cases}$ E. $f(t) = \begin{cases} \sin t, & \text{if } t > 0, \\ -\sin 2t, & \text{if } t \leqq 0. \end{cases}$

F. $f(t) = \begin{cases} (\sin t)^{1/5}, & \text{if } t < \dfrac{\pi}{2}, \\ -\cos t, & \text{if } t \geq \dfrac{\pi}{2}. \end{cases}$

【5】 Prove (3.7) [Any function $f(t)$ can be expressed as the sum of an even and an odd functions].

【6】 If a function $f(t)$ is odd, prove that $|f(t)|$ is even.

【7】 Calculate the coefficients of a Fourier series for each of the following periodic signals in **Figure 3.7**, where T, λ (not an integer), A, and ω_0 are constants:

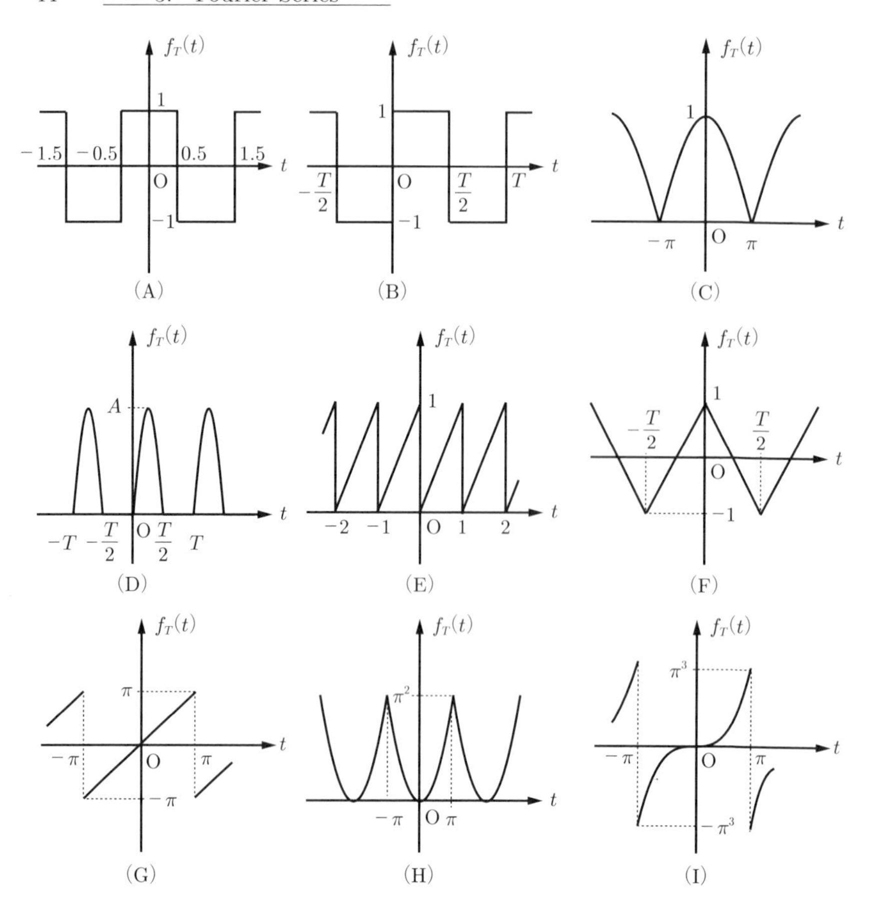

Figure 3.7 Signals for Basic-Level Problem 7

A. $$f_T(t) = \begin{cases} 1, & \text{if } 0 \leqq t < 0.5, \\ -1, & \text{if } 0.5 \leqq t < 1.5, \\ 1, & \text{if } 1.5 \leqq t < 2. \end{cases}$$

B. $$f_T(t) = \begin{cases} -1, & \text{if } -\dfrac{T}{2} \leqq t < 0, \\ 1, & \text{if } 0 \leqq t < \dfrac{T}{2}. \end{cases}$$

C. $f_T(t) = \cos \lambda t \quad \text{for } -\pi \leqq t < \pi.$

D. $$f_T(t) = \begin{cases} 0, & \text{if } -\dfrac{T}{2} \leqq t < 0, \\ A \sin \omega_0 t, & \text{if } 0 \leqq t < \dfrac{T}{2}. \end{cases}$$

E. $f_T(t) = t$ for $0 \leqq t < 1$.

F. $f_T(t) = \begin{cases} 1 + \dfrac{4t}{T}, & \text{if } -\dfrac{T}{2} \leqq t < 0, \\ 1 - \dfrac{4t}{T}, & \text{if } 0 \leqq t < \dfrac{T}{2}. \end{cases}$

G. $f_T(t) = t$ for $-\pi \leqq t < \pi$.

H. $f_T(t) = t^2$ for $-\pi \leqq t < \pi$.

I. $f_T(t) = t^3$ for $-\pi \leqq t < \pi$.

【8】 Following the derivation process of a_n given in Section 3.2, show the detailed derivation process of b_n.

【9】 Calculate the complex Fourier series of the periodic function in Basic-Lebel Problem 7 (c_n, $|c_n|$, and ϕ_n).

Advanced Level

【1】 Prove that $1 - \dfrac{1}{3} + \dfrac{1}{5} - \dfrac{1}{7} + \cdots = \dfrac{\pi}{4}$.

【2】 Solve the following problems:

A. Consider the Fourier series of Basic-Level Problems 7 G and H. Directly differentiate the functions and also find the Fourier series of their derivatives, $\mathrm{d}f_T(t)/\mathrm{d}t$. Check whether or not they are the same.

B. If a periodic function is continuous in $(-\infty, \infty)$ and satisfies a condition, then we can use the differentiation theorem to directly calculate the Fourier series of $\mathrm{d}f_T(t)/\mathrm{d}t$. Prove the following differentiation theorem for a Fourier series:

If $f_T(t)$ is continuous on $[-T/2, T/2]$ with $f_T(-T/2) = f_T(T/2)$ and $f_T(t+T) = f_T(t)$, and its derivative $\mathrm{d}f_T(t)/\mathrm{d}t$ is piecewise continuous and differentiable, then the Fourier series $f_T(t) = \dfrac{a_0}{2} + \displaystyle\sum_{n=1}^{\infty}(a_n \cos n\omega t + b_n \sin n\omega t)$ can be differentiated term-by-term and

$$\frac{\mathrm{d}f_T(t)}{\mathrm{d}t} = \sum_{n=1}^{\infty} n\omega(-a_n \sin n\omega t + b_n \cos n\omega t).$$

【3】 Consider a periodic function $f_T(t) = \mathrm{e}^t$ $(-\pi \leqq t < \pi)$ with period 2π, and let $\displaystyle\sum_{n=-\infty}^{\infty} c_n \mathrm{e}^{jn\omega t}$ be its Fourier series. Thus, $\mathrm{e}^t = \displaystyle\sum_{n=-\infty}^{\infty} c_n \mathrm{e}^{jn\omega t}$ holds for $|t| < \pi$. Formally differentiating both sides of this equation yields $\mathrm{e}^t =$

$\sum_{n=-\infty}^{\infty} jn\omega c_n \mathrm{e}^{jn\omega t}$. Thus, $c_n = jn\omega c_n$, or $(1 - jn\omega)c_n = 0$. As a result, $c_n = 0$ for all n. This is clearly not true. Find out what the problem is.

【4】 Prove the following theorem (integration of a Fourier series): Let $f_T(t)$ be piecewise continuous on $[-T/2, T/2]$ and $f_T(t+T) = f_T(t)$. If the Fourier series of $f_T(t)$ is $f_T(t) = \frac{a_0}{2} + \sum_{n=1}^{\infty}(a_n \cos n\omega t + b_n \sin n\omega t)$, then

$$\int_{t_1}^{t_2} f_T(t)\mathrm{d}t = \frac{a_0}{2}(t_2 - t_1)$$
$$+ \sum_{n=1}^{\infty} \frac{-b_n(\cos n\omega t_2 - \cos n\omega t_1) + a_n(\sin n\omega t_2 - \sin n\omega t_1)}{n\omega}.$$

【5】 Show that

A. $t^3 - \pi^2 t = 12 \sum_{n=1}^{\infty} \frac{(-1)^n \sin nt}{n^3} \quad (-\pi \leqq t \leqq \pi)$

B. $t^4 - 2\pi^2 t^2 = 48 \sum_{n=1}^{\infty} \frac{(-1)^{n+1} \cos nt}{n^4} - \frac{7\pi^4}{15} \quad (-\pi \leqq t \leqq \pi)$

C. $\sum_{n=1}^{\infty} \frac{1}{n^4} = \frac{\pi^4}{90}$.

4 Fourier Transform

As shown in the previous Chapter, a Fourier series is a powerful tool to deal with problems involving periodic signals. On the other hand, many signals are aperiodic. Thus, how to extend the Fourier series to handle aperiodic signals is important and practical. This chapter presents a solution to this problem.

4.1 Definition of Fourier Transform

The Fourier transform of an aperiodic function, $f(t)$, is defined to be

$$F(\omega) = \int_{-\infty}^{\infty} f(t)\mathrm{e}^{-j\omega t}\mathrm{d}t, \tag{4.1}$$

and its inverse Fourier transform is defined to be

$$f(t) = \frac{1}{2\pi}\int_{-\infty}^{\infty} F(\omega)\mathrm{e}^{j\omega t}\mathrm{d}\omega. \tag{4.2}$$

(4.1) and (4.2) are called a Fourier transform pair.

A Fourier transform, (4.1), exists if $f(t)$ satisfied the conditions in Section 3.2 (piecewise continuous and piecewise smooth) and is absolutely integrable, that is,

$$\int_{-\infty}^{\infty} |f(t)|\mathrm{d}t < \infty. \tag{4.3}$$

A function has two modes: $f(t)$ in the time domain and $F(\omega)$ in the frequency domain. (4.1) shows the frequency spectrum, $F(\omega)$, of $f(t)$, and (4.2) provides us with an inverse transform that collects its frequency spectrum to reconstruct the time-domain function, $f(t)$.

【Example 4.1】 Calculate the Fourier transform of an exponential decay formula (**Figure 4.1**) $f(t) = \begin{cases} 0, & \text{if } t < 0, \\ \mathrm{e}^{-\beta t}, & \text{if } t \geqq 0, \end{cases}$ where β is a positive constant.

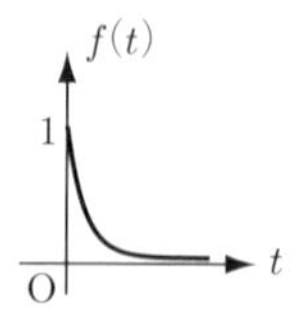

Figure 4.1 Exponential decay formula

It follows from (4.2) that

$$F(\omega) = \int_{-\infty}^{\infty} f(t)\mathrm{e}^{-j\omega t}\mathrm{d}t = \int_{0}^{\infty} \mathrm{e}^{-\beta t}\mathrm{e}^{-j\omega t}\mathrm{d}t = \int_{0}^{\infty} \mathrm{e}^{-(\beta+j\omega)t}\mathrm{d}t$$
$$= \frac{1}{\beta + j\omega}. \tag{4.4}$$

【Example 4.2】 Show that the Fourier transform of the Gaussian function $f(t) = \mathrm{e}^{-t^2}$ (**Figure 4.2**) is $F(\omega) = \sqrt{\pi}\mathrm{e}^{-\omega^2/4}$.

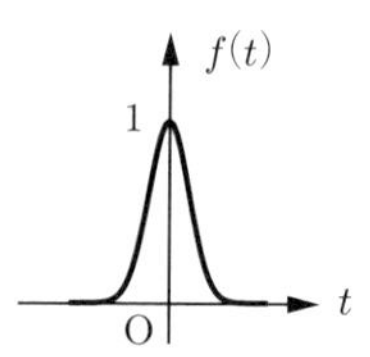

Figure 4.2 Gaussian function

It follows from (4.2) that

$$F(\omega) = \int_{-\infty}^{\infty} f(t)\mathrm{e}^{-j\omega t}\mathrm{d}t = \int_{-\infty}^{\infty} \mathrm{e}^{-t^2}\mathrm{e}^{-j\omega t}\mathrm{d}t = \int_{-\infty}^{\infty} \mathrm{e}^{-(t^2+j\omega t)}\mathrm{d}t. \tag{4.5}$$

Completing the square of the quadratic expression of t yields

$$t^2 + j\omega t = \left(t + \frac{j\omega}{2}\right)^2 + \frac{\omega^2}{4}. \tag{4.6}$$

Thus,

$$F(\omega) = \int_{-\infty}^{\infty} \mathrm{e}^{-\left[(t+j\omega/2)^2+\omega^2/4\right]}\mathrm{d}t = \mathrm{e}^{-\omega^2/4}\int_{-\infty}^{\infty} \mathrm{e}^{-(t+j\omega/2)^2}\mathrm{d}t.$$

Let $x = t + j\omega/2$. Then, $\mathrm{d}x = \mathrm{d}t$. Using (2.29) yields

$$F(\omega) = \mathrm{e}^{-\omega^2/4}\int_{-\infty}^{\infty} \mathrm{e}^{-x^2}\mathrm{d}x = \sqrt{\pi}\mathrm{e}^{-\omega^2/4}. \tag{4.7}$$

Note that the Fourier series of a periodic function is

$$\begin{cases} f_T(t) = \displaystyle\sum_{n=-\infty}^{\infty} c_n \mathrm{e}^{jn\omega t}, \\ c_n = \dfrac{1}{T}\displaystyle\int_{-T/2}^{T/2} f_T(t)\mathrm{e}^{-jn\omega t}\mathrm{d}t, \quad n = 0, \pm 1, \pm 2, \ldots. \end{cases} \tag{4.8}$$

Thus,

$$\begin{aligned} f_T(t) &= \sum_{n=-\infty}^{\infty} c_n \mathrm{e}^{jn\omega t} \\ &= \sum_{n=-\infty}^{\infty} \left[\frac{1}{T}\int_{-T/2}^{T/2} f_T(\tau)\mathrm{e}^{-jn\omega\tau}\mathrm{d}\tau\right]\mathrm{e}^{jn\omega t} \\ &= \frac{1}{T}\sum_{n=-\infty}^{\infty} \left[\int_{-T/2}^{T/2} f_T(\tau)\mathrm{e}^{-jn\omega\tau}\mathrm{d}\tau\right]\mathrm{e}^{jn\omega t}. \end{aligned} \tag{4.9}$$

If we let $\omega_n = n\omega$ and $\Delta\omega_n = \omega_{n+1} - \omega_n = \omega = 2\pi/T$, then

$$\begin{aligned} f_T(t) &= \frac{1}{2\pi}\sum_{n=-\infty}^{\infty} \left[\int_{-T/2}^{T/2} f_T(\tau)\mathrm{e}^{-jn\omega\tau}\mathrm{d}\tau\right]\mathrm{e}^{jn\omega t}\frac{2\pi}{T} \\ &= \frac{1}{2\pi}\sum_{n=-\infty}^{\infty} \left[\int_{-T/2}^{T/2} f_T(\tau)\mathrm{e}^{-j\omega_n\tau}\mathrm{d}\tau\right]\mathrm{e}^{j\omega_n t}\Delta\omega_n. \end{aligned} \tag{4.10}$$

Moreover, letting

$$F_T(\omega_n) = \int_{-T/2}^{T/2} f_T(\tau)\mathrm{e}^{-j\omega_n\tau}\mathrm{d}\tau \tag{4.11}$$

yields

$$f_T(t) = \frac{1}{2\pi}\sum_{n=-\infty}^{\infty} F_T(\omega_n)\mathrm{e}^{j\omega_n t}\Delta\omega_n. \tag{4.12}$$

Clearly, (4.1) and (4.2) are the limits of $f_T(t)$ and $F_T(\omega_n)$, respectively, as T approaches infinity.

Since an aperiodic function can be treated as a periodic function with its period being infinity. (4.1) and (4.2) are the extensions of c_n in (4.8) (except for the coefficient $1/T$) and (4.10), respectively.

4.2 Properties of Fourier Transform

The Fourier transform of a function $f(x)$, (4.1), is indicated to be†

$$F(\omega) = \mathcal{F}[f(t)]. \tag{4.13}$$

The transform has outstanding properties. Some of them are discussed below.

Linear transformation (linear mapping, linear function): Let the sets of variables and functions be X and Y, respectively; and the relationship between an item x in X ($x \in X$) and an item y in Y ($y \in Y$) be given by a function $y = \mathcal{T}(x)$. This is called a transformation from X to Y, and is indicated as $\mathcal{T} : X \to Y$.

A transformation is called a linear transformation if it satisfies the following conditions:

1) *Additivity* (加法性): $\mathcal{T}(u+v) = \mathcal{T}(u) + \mathcal{T}(v)$, $u, v \in X$.

† A is indicated to be B：A は B と表示される。

2) *Homogeneity of degree one* (斉一次性): $\mathcal{T}(au) = a\mathcal{T}(u)$, $u \in X$, $a \in \mathbb{R}$[†1].

Let $F(\omega) = \mathcal{F}[f(t)]$ and $G(\omega) = \mathcal{F}[g(t)]$ be the Fourier transforms of $f(t)$ and $g(t)$, respectively; $\mathcal{F}^{-1}[F(\omega)] = f(t)$ be the inverse Fourier transforms of $f(t)$; a and b be constants; and E be the energy content of $f(t)$. The properties of a Fourier transform are listed below.

Linearity: $\mathcal{F}[af(t) + bg(t)] = a\mathcal{F}[f(t)] + b\mathcal{F}[g(t)] = aF(\omega) + bG(\omega)$.

Proof: Calculating the required Fourier transform yields

$$\begin{aligned}\mathcal{F}[af(t) + bg(t)] &= \int_{-\infty}^{\infty} [af(t) + bg(t)]\mathrm{e}^{-j\omega t}\mathrm{d}t \\ &= a\int_{-\infty}^{\infty} f(t)\mathrm{e}^{-j\omega t}\mathrm{d}t + b\int_{-\infty}^{\infty} g(t)\mathrm{e}^{-j\omega t}\mathrm{d}t \\ &= aF(\omega) + bG(\omega). \end{aligned} \tag{4.14}$$

∎[†2]

Translation (time shifting): $\mathcal{F}[f(t \pm t_0)] = \mathrm{e}^{\pm j\omega t_0}\mathcal{F}[f(t)] = \mathrm{e}^{\pm j\omega t_0}F(\omega)$.

Proof: The definition of a Fourier transform gives

$$\mathcal{F}[f(t \pm t_0)] = \int_{-\infty}^{\infty} f(t \pm t_0)\mathrm{e}^{-j\omega t}\mathrm{d}t. \tag{4.15}$$

Letting $\tau = t \pm t_0$ yields $\mathrm{d}\tau = \mathrm{d}t$. Hence,

$$\begin{aligned}\int_{-\infty}^{\infty} f(t \pm t_0)\mathrm{e}^{-j\omega t}\mathrm{d}t &= \int_{-\infty}^{\infty} f(\tau)\mathrm{e}^{-j\omega(\tau \mp t_0)}\mathrm{d}\tau \\ &= \mathrm{e}^{\pm j\omega t_0}\int_{-\infty}^{\infty} f(\tau)\mathrm{e}^{-j\omega\tau}\mathrm{d}\tau \\ &= \mathrm{e}^{\pm j\omega t_0}F(\omega). \end{aligned} \tag{4.16}$$

∎

Translation (frequency shifting): $\mathcal{F}[f(t)\mathrm{e}^{j\omega_0 t}] = F(\omega - \omega_0)$.

Proof: The definition of a Fourier transform gives

---

†1 $\mathbb{R}$ is the set of real numbers

†2 ∎ is a symbol that is used at the end of a mathematical proof and means that a proof is completed. It is used in place of Q.E.D. for the Latin *phase quod erat demonstrandum*.

$$\mathcal{F}[f(t)\mathrm{e}^{j\omega_0 t}] = \int_{-\infty}^{\infty} f(t)\mathrm{e}^{j\omega_0 t}\mathrm{e}^{-j\omega t}\mathrm{d}t = \int_{-\infty}^{\infty} f(t)\mathrm{e}^{-j(\omega-\omega_0)t}\mathrm{d}t$$
$$= F(\omega-\omega_0). \tag{4.17}$$

∎

Derivative: $\mathcal{F}\left[\frac{\mathrm{d}f(t)}{\mathrm{d}t}\right] = j\omega\mathcal{F}[f(t)] = j\omega F(\omega).$

Proof: Differentiating both sides of the inverse Fourier transform of $f(t)$, (4.2), with time yields

$$\begin{aligned}\frac{\mathrm{d}f(t)}{\mathrm{d}t} &= \frac{1}{2\pi}\int_{-\infty}^{\infty}\frac{\partial[F(\omega)\mathrm{e}^{j\omega t}]}{\partial t}\mathrm{d}\omega = \frac{1}{2\pi}\int_{-\infty}^{\infty}F(\omega)\frac{\partial \mathrm{e}^{j\omega t}}{\partial t}\mathrm{d}\omega \\ &= \frac{1}{2\pi}\int_{-\infty}^{\infty}F(\omega)(j\omega\mathrm{e}^{j\omega t})\mathrm{d}\omega = \frac{1}{2\pi}\int_{-\infty}^{\infty}[j\omega F(\omega)]\mathrm{e}^{j\omega t}\mathrm{d}\omega \\ &= \frac{1}{2\pi}\int_{-\infty}^{\infty}[j\omega F(\omega)]\mathrm{e}^{j\omega t}\mathrm{d}\omega \\ &= \mathcal{F}^{-1}\left[j\omega F(\omega)\right].\end{aligned} \tag{4.18}$$

Thus,

$$\mathcal{F}\left[\frac{\mathrm{d}f(t)}{\mathrm{d}t}\right] = j\omega F(\omega). \tag{4.19}$$

∎

Integration: $\mathcal{F}\left[\int_{-\infty}^{t} f(t)\mathrm{d}t\right] = \frac{1}{j\omega}\mathcal{F}[f(t)] = \frac{1}{j\omega}F(\omega).$

Proof: Consider a function

$$g(t) = \int_{-\infty}^{t} f(\tau)\mathrm{d}\tau. \tag{4.20}$$

Since $\mathrm{d}g(t)/\mathrm{d}t = f(t)$, using the derivative property gives

$$\mathcal{F}\left[\frac{\mathrm{d}g(t)}{\mathrm{d}t}\right] = j\omega\mathcal{F}\left[g(t)\right]. \tag{4.21}$$

Thus,

$$\begin{aligned}\mathcal{F}\left[\int_{-\infty}^{t} f(\tau)\mathrm{d}\tau\right] &= \mathcal{F}\left[g(t)\right] = \frac{1}{j\omega}\mathcal{F}\left[\frac{\mathrm{d}g(t)}{\mathrm{d}t}\right] = \frac{1}{j\omega}\mathcal{F}[f(t)] \\ &= \frac{1}{j\omega}F(\omega).\end{aligned} \tag{4.22}$$

∎

Convolution (畳み込み積分) (**Time domain**):

$$\mathcal{F}\left[\int_{-\infty}^{\infty} f(\tau)g(t-\tau)\mathrm{d}\tau\right] = \mathcal{F}[f(t)*g(t)] = F(\omega)G(\omega). \tag{4.23}$$

Proof: According to the definition of a Fourier transform,

$$\begin{aligned}\mathcal{F}[f(t)*g(t)] &= \mathcal{F}\left[\int_{-\infty}^{\infty} f(\tau)g(t-\tau)\mathrm{d}\tau\right] \\ &= \int_{-\infty}^{\infty}\left[\int_{-\infty}^{\infty} f(\tau)g(t-\tau)\mathrm{d}\tau\right]\mathrm{e}^{-j\omega t}\mathrm{d}t. \end{aligned}\tag{4.24}$$

Letting $t-\tau=u$ yields

$$\begin{aligned}&\int_{-\infty}^{\infty}\left[\int_{-\infty}^{\infty} f(\tau)g(t-\tau)\mathrm{d}\tau\right]\mathrm{e}^{-j\omega t}\mathrm{d}t \\ &= \int_{-\infty}^{\infty}\int_{-\infty}^{\infty} f(\tau)\mathrm{e}^{-j\omega\tau}g(u)\mathrm{e}^{-j\omega u}\mathrm{d}\tau\mathrm{d}u \\ &= \int_{-\infty}^{\infty} f(\tau)\mathrm{e}^{-j\omega\tau}\mathrm{d}\tau\int_{-\infty}^{\infty} g(u)\mathrm{e}^{-j\omega u}\mathrm{d}u \\ &= F(\omega)G(\omega). \end{aligned}\tag{4.25}$$

∎

Convolution (**Frequency domain**): $\mathcal{F}^{-1}\left[F(\omega)*G(\omega)\right] = 2\pi f(t)g(t)$ or $\mathcal{F}\left[f(t)g(t)\right] = \dfrac{1}{2\pi}F(\omega)*G(\omega) = \dfrac{1}{2\pi}\displaystyle\int_{-\infty}^{\infty}F(\sigma)G(\omega-\sigma)\mathrm{d}\sigma$

Proof: According to the definition of an inverse Fourier transform,

$$\mathcal{F}^{-1}[F(\omega)*G(\omega)] = \frac{1}{2\pi}\int_{-\infty}^{\infty}\left[\int_{-\infty}^{\infty} F(\sigma)G(\omega-\sigma)\mathrm{d}\sigma\right]\mathrm{e}^{j\omega t}\mathrm{d}\omega. \tag{4.26}$$

Letting $\omega-\sigma=\chi$ yields

$$\begin{aligned}&\frac{1}{2\pi}\int_{-\infty}^{\infty}\left[\int_{-\infty}^{\infty} F(\sigma)G(\omega-\sigma)\mathrm{d}\sigma\right]\mathrm{e}^{j\omega t}\mathrm{d}\omega \\ &= \frac{1}{2\pi}\int_{-\infty}^{\infty}\int_{-\infty}^{\infty} F(\sigma)\mathrm{e}^{j\sigma t}\left[G(\chi)\mathrm{e}^{j\chi t}\right]\mathrm{d}\chi\mathrm{d}\sigma \\ &= 2\pi\left[\frac{1}{2\pi}\int_{-\infty}^{\infty} F(\sigma)\mathrm{e}^{j\sigma t}\mathrm{d}\sigma\right]\left[\frac{1}{2\pi}\int_{-\infty}^{\infty} G(\chi)\mathrm{e}^{j\chi t}\mathrm{d}\chi\right] \\ &= 2\pi f(t)g(t). \end{aligned}\tag{4.27}$$

∎

Parseval's identity: $E = \int_{-\infty}^{\infty} |f(t)|^2 \mathrm{d}t = \frac{1}{2\pi} \int_{-\infty}^{\infty} |F(\omega)|^2 \mathrm{d}\omega.$

Proof: Letting the complex conjugate transpose of $f(t)$ be $f^*(t)$, we have

$$\begin{aligned}
\mathcal{F}[f^*(t)] &= \int_{-\infty}^{\infty} f^*(t)\mathrm{e}^{-j\omega t}\mathrm{d}t = \int_{-\infty}^{\infty} \left[f(t)\mathrm{e}^{j\omega t}\right]^* \mathrm{d}t \\
&= \left[\int_{-\infty}^{\infty} f(t)\mathrm{e}^{-j(-\omega)t}\mathrm{d}t\right]^* = F^*(-\omega).
\end{aligned} \tag{4.28}$$

Thus,

$$\begin{aligned}
E &= \int_{-\infty}^{\infty} |f(t)|^2 \mathrm{d}t = \int_{-\infty}^{\infty} f^*(t)f(t)\mathrm{d}t \\
&= \frac{1}{2\pi}\int_{-\infty}^{\infty} F^*[-(-\omega)]F(\omega)\mathrm{d}\omega = \frac{1}{2\pi}\int_{-\infty}^{\infty} F^*(\omega)F(\omega)\mathrm{d}\omega \\
&= \frac{1}{2\pi}\int_{-\infty}^{\infty} |F(\omega)|^2 \mathrm{d}\omega.
\end{aligned} \tag{4.29}$$

∎

4.3 Spectrum, Energy Spectral Density, and Correlation Function

As explained in Section 3.3, the spectrum of a periodic signal, $|c_n|$, in (3.64) describes the distribution of the amplitude of each harmonic with respect to frequency. It is called a discrete spectrum due to its discontinuity. The spectrum clearly shows frequency components and the proportion of each component in a periodic function.

For an aperiodic signal $f(t)$, its Fourier transform $F(\omega)$ is called the spectrum or spectral function of the signal. The amplitude spectrum (or simply spectrum) $|F(\omega)|$ is an even function

$$|F(-\omega)| = |F(\omega)| \tag{4.30}$$

and the phase spectrum is an odd function

$$\angle F(-\omega) = -\angle F(\omega), \quad \angle F(\omega) = \tan^{-1}\frac{\mathrm{Im}[F(\omega)]}{\mathrm{Re}[F(\omega)]}. \tag{4.31}$$

We call

$$\mathrm{ESD}(\omega) = |F(\omega)|^2 \tag{4.32}$$

the *energy spectral density* (エネルギースペクトル密度) of $f(t)$, which determines the energy distribution of $f(t)$ over frequency.

【Example 4.3】 Calculate the spectra of a periodic function (**Figure 4.3**)

$$f_T(t) = \begin{cases} 0, & \text{if } -\dfrac{T}{2} \leqq t < -\dfrac{\tau}{2}, \\ A, & \text{if } -\dfrac{\tau}{2} \leqq t < \dfrac{\tau}{2}, \\ 0, & \text{if } \dfrac{\tau}{2} \leqq t < \dfrac{T}{2} \end{cases} \tag{4.33}$$

and an aperiodic function

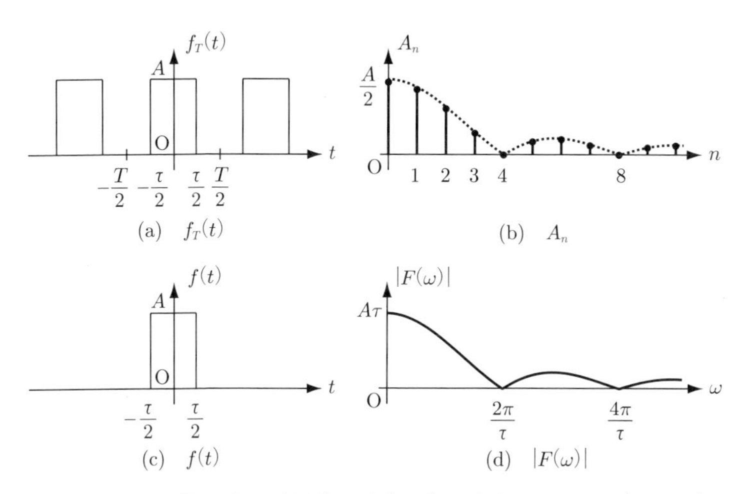

Figure 4.3 Functions (4.33) and (4.34) and their spectra ($T = 4\tau$)

$$f(t)=\begin{cases}0, & -\infty<t<-\dfrac{\tau}{2},\\ A, & -\dfrac{\tau}{2}\leq t\leq\dfrac{\tau}{2},\\ 0, & \dfrac{\tau}{2}<t<\infty\end{cases} \tag{4.34}$$

and compare them, where T is the period and τ ($\leq T$) is a constant (Figure 4.3).

The Fourier series of $f_T(t)$ is

$$f_T(t)=\frac{A\tau}{T}+\sum_{n=-\infty,n\neq 0}^{\infty}\frac{A}{n\pi}\sin\frac{n\pi\tau}{T}\mathrm{e}^{jn\omega_0 t},\ \omega_0=\frac{2\pi}{T}. \tag{4.35}$$

The coefficients of the complex Fourier series are

$$c_0=\frac{A\tau}{T},\ c_n=\frac{A}{n\pi}\sin\frac{n\pi\tau}{T},n=\pm1,\pm2,\ldots. \tag{4.36}$$

Thus, its single-sided spectrum (see the explain after Example 3.3) is

$$\begin{cases}A_0=2|c_0|=\dfrac{2A\tau}{T},\\ A_n=2|c_n|=\dfrac{2A}{n\pi}\left|\sin\dfrac{n\pi\tau}{T}\right|,n=1,2,\ldots.\end{cases} \tag{4.37}$$

On the other hand,

$$F(\omega)=\int_{-\infty}^{\infty}f(t)\mathrm{e}^{-j\omega t}\mathrm{d}t=\int_{-\tau/2}^{\tau/2}A\mathrm{e}^{-j\omega t}\mathrm{d}t=\frac{2A}{\omega}\sin\frac{\omega\tau}{2}. \tag{4.38}$$

Thus, its spectrum is

$$|F(\omega)|=2A\left|\frac{\sin\dfrac{\omega\tau}{2}}{\omega}\right|. \tag{4.39}$$

Figure 4.3 shows the signals in the time domain and their single-sided spectra ($\omega\geq 0$). A comparison of these two spectra shows that they have the same shape. While the spectrum of the periodic function $f_T(t)$ is a discrete one and takes values at the DC (A_0), and the fundamental

(A_1) and harmonic frequencies ($A_2, A_3, \ldots$), that of the aperiodic function $f(t)$ is a continuous one over all frequencies $\omega \in [0, \infty)$.

Correlation functions (相関関数) are also an important concept in spectrum analysis. For two functions $f_1(t)$ and $f_2(t)$,

$$\begin{aligned} R_{12}(\tau) &= \int_{-\infty}^{\infty} f_1(t) f_2(t+\tau) \mathrm{d}t, \\ R_{21}(\tau) &= \int_{-\infty}^{\infty} f_1(t+\tau) f_2(t) \mathrm{d}t \end{aligned} \tag{4.40}$$

are called the *cross-correlation functions* (相互相関関数) of $f_1(t)$ and $f_2(t)$. When $f_1(t) = f_2(t) = f(t)$,

$$R(\tau) = \int_{-\infty}^{\infty} f(t) f(t+\tau) \mathrm{d}t \tag{4.41}$$

is called an *autocorrelative function* (or simply *correlative function*) (自己相関関数) of $f(t)$.

Since

$$R(-\tau) = \int_{-\infty}^{\infty} f(t) f(t-\tau) \mathrm{d}t,$$

defining $t = v + \tau$ yields

$$R(-\tau) = \int_{-\infty}^{\infty} f(v+\tau) f(v) \mathrm{d}v = R(\tau). \tag{4.42}$$

Thus, $R(\tau)$ is an even function. In the same manner, we can easily prove that

$$R_{21}(\tau) = R_{12}(-\tau). \tag{4.43}$$

Verification shows that $\mathrm{ESD}(\omega)$ and $R(\tau)$ are a pair of a Fourier transform, that is,

$$\begin{cases} \mathrm{ESD}(\omega) = \displaystyle\int_{-\infty}^{\infty} R(\tau) \mathrm{e}^{-j\omega\tau} \mathrm{d}\tau, \\ R(\tau) = \dfrac{1}{2\pi} \displaystyle\int_{-\infty}^{\infty} \mathrm{ESD}(\omega) \mathrm{e}^{j\omega\tau} \mathrm{d}\omega. \end{cases} \tag{4.44}$$

Note that both $\mathrm{ESD}(\omega)$ and $R(\tau)$ are even functions. We can write (4.44) as

$$\begin{cases} \mathrm{ESD}(\omega) = \displaystyle\int_{-\infty}^{\infty} R(\tau)\cos\omega\tau \mathrm{d}\tau, \\ R(\tau) = \dfrac{1}{2\pi}\displaystyle\int_{-\infty}^{\infty} \mathrm{ESD}(\omega)\cos\omega\tau \mathrm{d}\omega. \end{cases} \tag{4.45}$$

When $\tau = 0$,

$$R(0) = \int_{-\infty}^{\infty} [f(t)]^2 \mathrm{d}t = \frac{1}{2\pi}\int_{-\infty}^{\infty} \mathrm{ESD}(\omega)\mathrm{d}\omega. \tag{4.46}$$

It is exactly Parseval's identity.

Let $F_1(\omega) = \mathcal{F}[f_1(t)]$ and $F_2(\omega) = \mathcal{F}[f_2(t)]$. The cross-energy density spectrum of $f_1(t)$ and $f_2(t)$ is defined to be

$$S_{12}(\omega) = F_1^*(\omega)F_2(\omega), \tag{4.47}$$

where $F_1^*(\omega)$ is the complex conjugate of $F_1(\omega)$. A verification shows that $S_{12}(\omega)$ and $R_{12}(\tau)$ are also a pair of a Fourier transform:

$$\begin{cases} S_{12}(\omega) = \displaystyle\int_{-\infty}^{\infty} R_{12}(\tau)\mathrm{e}^{-j\omega\tau}\mathrm{d}\tau, \\ R_{12}(\tau) = \dfrac{1}{2\pi}\displaystyle\int_{-\infty}^{\infty} S_{12}(\omega)\mathrm{e}^{j\omega\tau}\mathrm{d}\omega. \end{cases} \tag{4.48}$$

4.4 Fourier Transforms of Special Functions

A function, $f(t)$, is required to be *absolutely integrable* (絶対可積分), that is, $\int_{-\infty}^{\infty} |f(t)|\mathrm{d}t < \infty$, to ensure the existence of its Fourier transform, $\mathcal{F}(j\omega)$. Many well-known functions are not absolutely integrable, such as trigonometric functions and the step function. However, their Fourier transforms can be defined using the unit impulse function, $\delta(t)$, shown in Subsection 2.6.3.

The Fourier transform of the unit impulse function is

$$\mathcal{F}[\delta(t)] = \int_{-\infty}^{\infty} \delta(t)\mathrm{e}^{-j\omega t}\mathrm{d}t = \mathrm{e}^{-j\omega t}|_{t=0} = 1. \tag{4.49}$$

The Fourier transform of the unit impulse function indicates that the function contains all frequency components. Its inverse Fourier transform is

$$\delta(t) = \mathcal{F}^{-1}(1) = \frac{1}{2\pi}\int_{-\infty}^{\infty} 1\mathrm{e}^{j\omega t}\mathrm{d}\omega = \frac{1}{2\pi}\int_{-\infty}^{\infty} \mathrm{e}^{j\omega t}\mathrm{d}\omega. \tag{4.50}$$

Using Euler's formula, (2.3), yields

$$\begin{aligned}\delta(t) &= \frac{1}{2\pi}\int_{-\infty}^{\infty}(\cos\omega t + j\sin\omega t)\mathrm{d}\omega \\ &= \frac{1}{2\pi}\int_{-\infty}^{\infty}\cos\omega t\mathrm{d}\omega + j\frac{1}{2\pi}\int_{-\infty}^{\infty}\sin\omega t\mathrm{d}\omega \\ &= \frac{1}{\pi}\int_{0}^{\infty}\cos\omega t\mathrm{d}\omega. \end{aligned} \tag{4.51}$$

Note that we used the properties of even and odd functions for integration in the last equation.

Observing (4.50) shows that it is a mapping from the frequency domain (ω) to the time domain (t). On the other hand, if we switch the variables of t and ω, it provides us with a mapping from the time domain to the frequency domain:

$$\delta(\omega) = \frac{1}{2\pi}\int_{-\infty}^{\infty}\mathrm{e}^{jt\omega}\mathrm{d}t. \tag{4.52}$$

Thus, we yield

$$\delta(-\omega) = \frac{1}{2\pi}\int_{-\infty}^{\infty}\mathrm{e}^{jt(-\omega)}\mathrm{d}t = \frac{1}{2\pi}\int_{-\infty}^{\infty}\mathrm{e}^{-jt\omega}\mathrm{d}t. \tag{4.53}$$

Note that $\delta(-\omega) = \delta(\omega)$. We also have

$$\delta(\omega) = \frac{1}{2\pi}\int_{-\infty}^{\infty}\mathrm{e}^{-jt\omega}\mathrm{d}t. \tag{4.54}$$

(4.52), (4.53), and (4.54) play important roles in the calculations of special functions.

Now, we show the Fourier transform of a constant. Consider the following function:

$$f(t) = a. \tag{4.55}$$

Its Fourier transform is

$$\mathcal{F}(a) = \int_{-\infty}^{\infty} a\mathrm{e}^{-j\omega t}\mathrm{d}t = 2\pi a\left[\frac{1}{2\pi}\int_{-\infty}^{\infty} \mathrm{e}^{-j\omega t}\mathrm{d}t\right]. \tag{4.56}$$

It follows from (4.54) that

$$\mathcal{F}(a) = 2\pi a\delta(\omega). \tag{4.57}$$

Let us consider the Fourier transforms of $\cos\omega_0 t$ and $\sin\omega_0 t$. Euler's formula gives

$$\cos\omega_0 t = \frac{\mathrm{e}^{j\omega_0 t} + \mathrm{e}^{j\omega_0 t}}{2}, \tag{4.58}$$

$$\sin\omega_0 t = \frac{\mathrm{e}^{j\omega_0 t} - \mathrm{e}^{j\omega_0 t}}{2j}. \tag{4.59}$$

Thus, the Fourier transforms of $\cos\omega_0 t$ and $\sin\omega_0 t$ can be described by the combination of the Fourier transforms of $\mathrm{e}^{j\omega_0 t}$ and $\mathrm{e}^{-j\omega_0 t}$. Substituting $a = 1$ into (4.57) yields

$$\mathcal{F}(1) = 2\pi\delta(t). \tag{4.60}$$

The frequency-shifting property in Section 4.2 shows that $\mathcal{F}[f(t)\mathrm{e}^{j\omega_0 t}] = F(\omega - \omega_0)$. If we shift the constant 1 by $\mathrm{e}^{j\omega_0 t}$ and $\mathrm{e}^{j\omega_0(-t)}$, it gives

$$\mathcal{F}(\mathrm{e}^{j\omega_0 t}) = 2\pi\delta(\omega - \omega_0), \tag{4.61}$$

$$\mathcal{F}(\mathrm{e}^{-j\omega_0 t}) = 2\pi\delta(\omega + \omega_0). \tag{4.62}$$

Therefore, the Fourier transforms of $\cos\omega_0 t$ and $\sin\omega_0 t$ are

$$\begin{aligned}\mathcal{F}(\cos\omega_0 t) &= \frac{\mathcal{F}(\mathrm{e}^{j\omega_0 t}) + \mathcal{F}(\mathrm{e}^{-j\omega_0 t})}{2} \\ &= \pi[\delta(\omega - \omega_0) + \delta(\omega + \omega_0)],\end{aligned} \tag{4.63}$$

$$\begin{aligned}\mathcal{F}(\sin \omega_0 t) &= \frac{\mathcal{F}(\mathrm{e}^{j\omega_0 t}) - \mathcal{F}(\mathrm{e}^{-j\omega_0 t})}{2j} \\ &= -j\pi[\delta(\omega - \omega_0) - \delta(\omega + \omega_0)]. \end{aligned} \tag{4.64}$$

[Example 4.4] Calculate the Fourier transform of the unit step function (2.37): $1(t) = \begin{cases} 0, & \text{if } t < 0, \\ 1, & \text{if } t \geqq 0. \end{cases}$

It follows from (2.37) that

$$1(-t) = \begin{cases} 0, & \text{if } t > 0, \\ 1, & \text{if } t \leqq 0. \end{cases} \tag{4.65}$$

Let $\mathcal{F}[1(t)] = F(\omega)$. It is true that $\mathcal{F}[1(-t)] = F(-\omega)$ (See Basic-Level Problem 9 in this chapter). Since $1(t) + 1(-t) = 1$ holds excpet at $t = 0$, $\mathcal{F}[1(t)] + \mathcal{F}[1(-t)] = \mathcal{F}(1)$, that is,

$$F(\omega) + F(-\omega) = 2\pi\delta(\omega). \tag{4.66}$$

Assume that

$$F(\omega) = k\delta(\omega) + P(\omega), \tag{4.67}$$

where k is a constant and $P(\omega)$ is an ordinary function. Note that $\delta(-\omega) = \delta(\omega)$. It follows from (4.67) that

$$\begin{aligned}F(\omega) + F(-\omega) &= k\delta(\omega) + P(\omega) + k\delta(-\omega) + P(-\omega) \\ &= 2k\delta(\omega) + P(\omega) + P(-\omega). \end{aligned} \tag{4.68}$$

A comparison between (4.66) and (4.68) reveals that

$$k = \pi, \; P(\omega) + P(-\omega) = 0. \tag{4.69}$$

Hence, $P(\omega)$ is an odd function.

On the other hand, $\mathrm{d}1(t)/\mathrm{d}t = \delta(t)$. The differential property (4.19) provides us with

$$\frac{\mathrm{d}1(t)}{\mathrm{d}t} = j\omega F(\omega) = j\omega[\pi\delta(\omega) + P(\omega)] = j\pi\omega\delta(\omega) + j\omega P(\omega). \tag{4.70}$$

Since $\mathcal{F}[\delta(t)] = 1$ and it is easy to show that $\omega\delta(\omega) = 0$ [see (2.49)],

$$P(\omega) = \frac{1}{j\omega}. \tag{4.71}$$

As a result, we obtain

$$\mathcal{F}[1(t)] = \pi\delta(\omega) + \frac{1}{j\omega}. \tag{4.72}$$

Note that the relationship $\omega F_1(\omega) = \omega F_2(\omega)$ does not provide us with $F_1(\omega) = F_2(\omega)$ but $F_1(\omega) = F_2(\omega) + k\delta(\omega)$, where k is a constant, because $\omega\delta(\omega) = 0$.

Problems

Basic Level

【1】 Find the Fourier transforms of the following functions:

A. $f(t) = A(t)\cos\omega_0 t$, where $A(t) = \begin{cases} 1, & \text{if } |t| \leqq \dfrac{D}{2}, \\ 0, & \text{if } |t| > D. \end{cases}$

B. $f(t) = \begin{cases} 1 - t^2, & \text{if } t^2 < 1, \\ 0, & \text{if } t^2 \geqq 1. \end{cases}$ C. $f(t) = \begin{cases} 0, & \text{if } t < 0, \\ \mathrm{e}^{-t}\sin 2t, & \text{if } t \geqq 0. \end{cases}$

D. $f(t) = \begin{cases} 0, & \text{if } -\infty < t < -1, \\ -1, & \text{if } -1 \leqq t < 0, \\ 1, & \text{if } 0 \leqq t < 1, \\ 0, & \text{if } 1 \leqq t < \infty. \end{cases}$

E. $f(t) = \dfrac{1}{2}\left[\delta(t+a) + \delta(t-a) + \delta\left(t + \dfrac{a}{2}\right) + \delta\left(t - \dfrac{a}{2}\right)\right].$

【2】 Find the spectrum of each of the following functions (**Figure 4.4**):

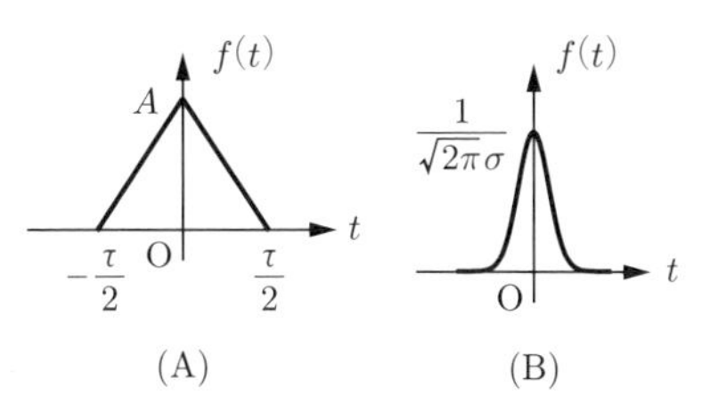

Figure 4.4 Basic-Level Problem 2

A. A triangular pulse: $f(t) = \begin{cases} 0, & \text{if } -\infty < t < -\dfrac{\tau}{2}, \\ A + \dfrac{2A}{\tau}t, & \text{if } -\dfrac{\tau}{2} \leqq t < 0, \\ A - \dfrac{2A}{\tau}t, & \text{if } 0 \leqq t < \dfrac{\tau}{2}, \\ 0, & \text{if } \dfrac{\tau}{2} \leqq t < \infty. \end{cases}$

B. A normal distribution (Gaussian distribution):
$f(t) = \dfrac{1}{\sqrt{2\pi}\sigma} \mathrm{e}^{-\frac{t^2}{2\sigma^2}}$.

【3】 Prove that, if $f_T(t)$ is a non-sinusoidal periodic function with period T, then its spectrum is $F_T(\omega) = 2\pi \displaystyle\sum_{n=-\infty}^{\infty} c_n \delta(\omega - n\omega_0)$, where $\omega_0 = 2\pi/T$ and c_n are the coefficients of the complex Fourier series of $f_T(t)$.

【4】 Solve the following problems:

A. Calculate the Fourier transform of $f(t) = \mathrm{e}^{-a|t|}$ $(a \neq 0)$ and prove that $\displaystyle\int_0^\infty \frac{\cos \omega t}{a^2 + \omega^2} \mathrm{d}\omega = \frac{\pi}{2a} \mathrm{e}^{-a|t|}$.

B. Calculate the Fourier transform of $f(t) = \mathrm{e}^{-|t|} \cos t$ and prove that $\displaystyle\int_0^\infty \frac{\omega^2 + 2}{\omega^2 + 4} \cos \omega t \mathrm{d}\omega = \frac{\pi}{2} \mathrm{e}^{-|t|} \cos t$.

C. Calculate the Fourier transform of $f(t) = \begin{cases} \sin t, & \text{if } |t| \leqq \pi, \\ 0, & \text{if } |t| > \pi. \end{cases}$ and prove that $\displaystyle\int_0^\infty \frac{\sin \omega\pi \sin \omega t}{1 - \omega^2} \mathrm{d}\omega = \begin{cases} \dfrac{\pi}{2} \sin t, & \text{if } |t| \leqq \pi, \\ 0, & \text{if } |t| > \pi \end{cases}$.

【5】 Let $F(\omega) = \mathcal{F}[f(t)]$ and a is a nonzero real number. Show that $\mathcal{F}[f(at)] = \dfrac{1}{|a|} F\left(\dfrac{\omega}{a}\right)$.

【6】 Let $F(\omega) = \mathcal{F}[f(t)]$. Prove that $\dfrac{\mathrm{d}F(\omega)}{\mathrm{d}\omega} = -j\mathcal{F}[tf(t)]$.

【7】 Calculate $\mathcal{F}\left[\dfrac{\mathrm{d}^n f(t)}{\mathrm{d}t^n}\right]$.

【8】 Find $f(t)$ for given $F(s)$:

A. $F(\omega) = \dfrac{\sin\omega}{\omega}$, B. $F(s) = \dfrac{1}{a + j\omega}$.

【9】 Prove that, if $F(\omega) = \mathcal{F}[f(t)]$, then $F(-\omega) = \mathcal{F}[f(-t)]$.

【10】 Prove that, if $F(\omega) = \mathcal{F}[f(t)]$, then

$\mathcal{F}[f(t)\cos\omega_0 t] = \dfrac{F(\omega - \omega_0) + F(\omega + \omega_0)}{2}$ and

$\mathcal{F}[f(t)\sin\omega_0 t] = \dfrac{F(\omega - \omega_0) + F(\omega - \omega_0)}{2j}$.

【11】 Prove that, if $F(\omega) = \mathcal{F}[f(t)]$, then $f(\pm\omega) = \dfrac{1}{2\pi}\displaystyle\int_{-\infty}^{\infty} F(\mp t)\mathrm{e}^{-j\omega t}\mathrm{d}t$.

【12】 Prove (4.30) and (4.31).

Advanced Level

【1】 Show that the unit step function $1(t)$ can be expressed as $1(t) = \dfrac{1}{2} + \dfrac{1}{\pi}\displaystyle\int_0^{\infty} \dfrac{\sin\omega t}{\omega}\mathrm{d}\omega$.

【2】 Show that $\mathcal{F}\left(\dfrac{1}{t^4 + 1}\right) = \dfrac{\pi}{\sqrt{2}}\mathrm{e}^{-\frac{|\omega|}{\sqrt{2}}}\left(\cos\dfrac{\omega}{\sqrt{2}} + \sin\dfrac{\omega}{\sqrt{2}}\right)$.

【3】 Prove (4.44).

【4】 Prove (4.48).

【5】 Prove that

A. $f_1(t) * f_2(t) = f_2(t) * f_1(t)$.

B. $f_1(t) * [f_2(t) * f_3(t)] = [f_1(t) * f_2(t)] * f_3(t)$.

C. $\mathrm{e}^{at}[f_1(t) * f_2(t)] = [\mathrm{e}^{at} f_1(t)] * [\mathrm{e}^{at} f_2(t)]$ (a: constant).

D. $\dfrac{\mathrm{d}[f_1(t) * f_2(t)]}{\mathrm{d}t} = \dfrac{\mathrm{d}f_1(t)}{\mathrm{d}t} * f_2(t) = f_1(t) * \dfrac{\mathrm{d}f_2(t)}{\mathrm{d}t}$.

5 Signal Sampling and Reconstruction

We usually use a computer to perform a Fourier and an inverse Fourier transforms. So, we have to sample a continuous signal $s(t)$ with a sampling period ΔT to produce a discrete sequence $\{s(k\Delta T)\}$ $(k = 0, 1, 2, \ldots)$. We usually simply write the sequence as $\{s[k]\}$ $(k = 0, 1, 2, \ldots)$. While a small ΔT produces a sequence that contains rich information of an original continuous signal and its Fourier transform is close to the continuous one, the amount of digital data is big and the Fourier and inverse Fourier transforms are computationally expense. How to choose a suitable sampling period to perform a Fourier and an inverse Fourier transforms is an important issue[†] in implementing Fourier analysis.

5.1 The Sampling Theorem

The *sampling theorem* (or uniform sampling theorem, Nyquist-Shannon sampling theorem) (サンプリング定理) specifies the maximum sampling period that permits a discrete sequence to capture all the information of a continuous signal so that the original one can be completely recovered or reconstructed. It is one of the most important theorems in the field of information theory and is stated as follows:

If a time-variant signal $s(t)$ does not have angular frequency components higher than ω_M [rad/s], then the signal can be completely determined if it is sampled with an angular sampling frequency higher than $2\omega_M$ [rad/s].

† $\sim$ is an important issue：～は重要課題（重要問題）である

Let a sampling angular frequency be ω_S. Half of it, $\omega_N = \omega_S/2$, is called the *Nyquist angular frequency* (ナイキスト角周波数). The sampling theorem states that the sampling angular frequency should be higher than $2\omega_M$ to recover all frequency components of the signal.

Let us use an example to understand the theorem. A signal

$$s(t) = \cos 2\pi t + \cos 4\pi t + \cos 8\pi t \tag{5.1}$$

and its Fourier transform are shown in **Figure 5.1**. The signal has three frequency components: 2π, 4π, and 8π rad/s. The highest angular frequency, ω_M, is 8π $(= 25.1)$ rad/s.

If we sample the signal at an angular frequency $\omega_S = 201$ rad/s ($> 2 \times 8\pi = 50.3$ rad/s), we see that the three frequency components of the signal are precisely acquired (**Figure 5.2**).

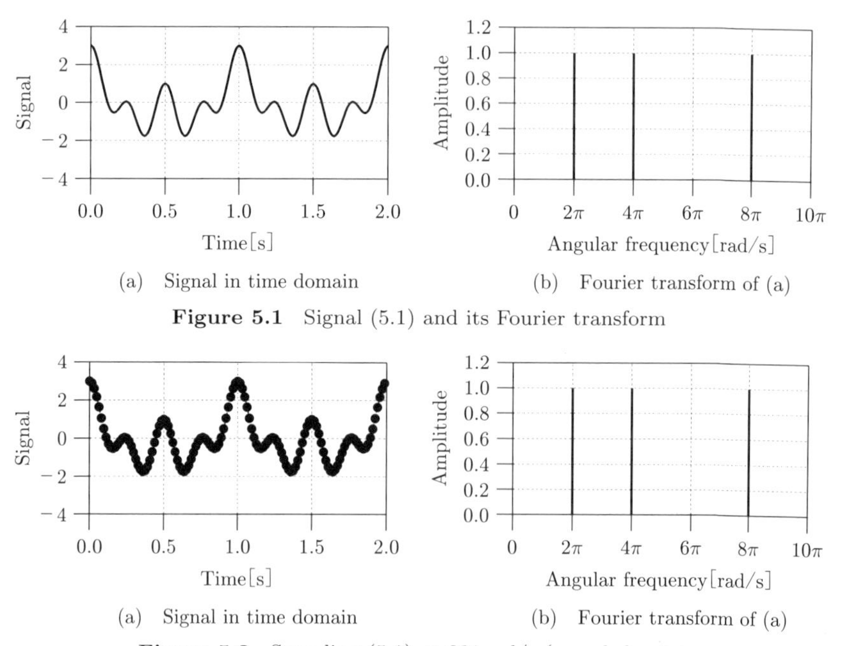

(a) Signal in time domain (b) Fourier transform of (a)

Figure 5.1 Signal (5.1) and its Fourier transform

(a) Signal in time domain (b) Fourier transform of (a)

Figure 5.2 Sampling (5.1) at 201 rad/s (sampled points and its Fourier transform)

Then, we lower the sampling angular frequency to 101 rad/s ($>$ 50.3 rad/s) to reduce data size. **Figure 5.3** shows that the three frequency components of the signal are still precisely acquired.

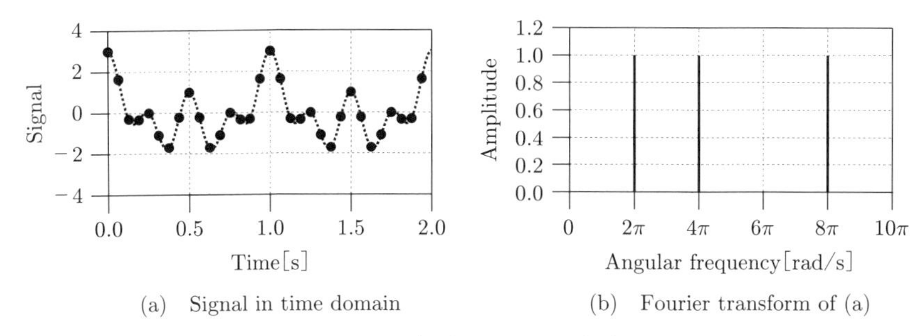

(a) Signal in time domain (b) Fourier transform of (a)

Figure 5.3 Sampling (5.1) at 101 rad/s (sampled points and its Fourier transform)

We further lower the sampling angular frequency to 50.3 rad/s, which is exactly twice ω_M. The Fourier transform result (**Figure 5.4**) shows that the frequency components are suitably acquired.

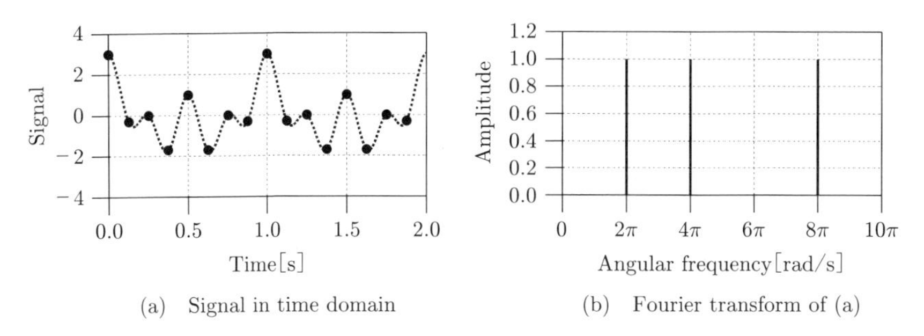

(a) Signal in time domain (b) Fourier transform of (a)

Figure 5.4 Sampling (5.1) at 50.3 rad/s (sampled points and its Fourier transform)

It is worth mentioning that† the sampling angular frequency $2\omega_M$ is sensitive. While the frequency is good for the signal (5.1), it fails to find the amplitude of the highest frequency (**Figure 5.5**) for a signal

† It is worth mentioning that ～：これは役に立つ（重要な）ことですが，～。

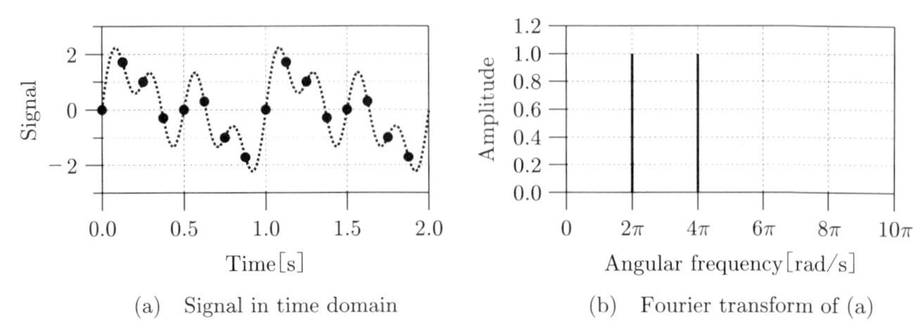

(a) Signal in time domain (b) Fourier transform of (a)

Figure 5.5 Sampling (5.2) at 50.3 rad/s (sampled points and its Fourier transform)

$$s(t) = \sin 2\pi t + \sin 4\pi t + \sin 8\pi t. \tag{5.2}$$

Note that the signal has the same frequency components as (5.1) does. This reveals that we need to use a sampling frequency that is higher than $2\omega_M$ to ensure that a signal can be properly transformed into the frequency domain.

On the other hand, if we choose a sampling angular frequency of $\omega_S = 25.1$ rad/s, which equals ω_M, the component of ω_M appears at the angular frequency $\omega = 0$ (**Figure 5.6**).

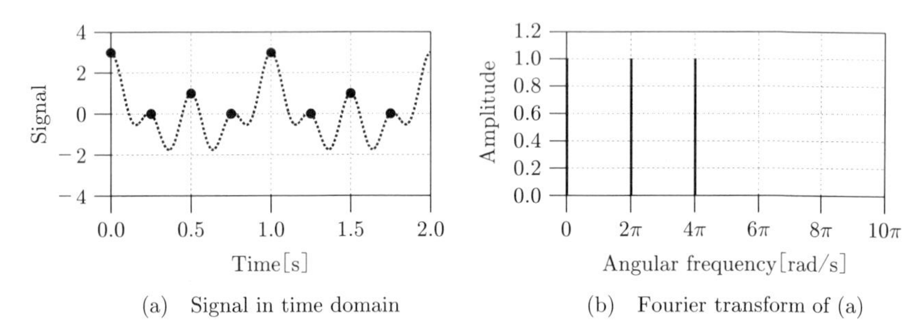

(a) Signal in time domain (b) Fourier transform of (a)

Figure 5.6 Sampling (5.1) at 25.1 rad/s (sampled points and its Fourier transform)

This phenomenon is called *aliasing* (エイリアシング). Generally speaking, when a sine curve with angular frequency ω_1 is sampled at an angular frequency ω_S, the samples are indistinguishable from samples of angular frequencies $\omega_n = \omega_1 - n\omega_S$, where n is an integer. More specifically, aliasing is a phenomenon in which angular-frequency components higher than $\omega_S/2$ (for example, ω_1) appear to be folded back[†] to the low-frequency side with $\omega_S/2$ as the center (it is called the folding angular frequency), that is, at the angular frequency (**Figure 5.7**)

$$\omega_1' = \omega_1 - 2\Delta = \omega_1 - 2\left(\omega_1 - \frac{\omega_S}{2}\right) = \omega_S - \omega_1. \tag{5.3}$$

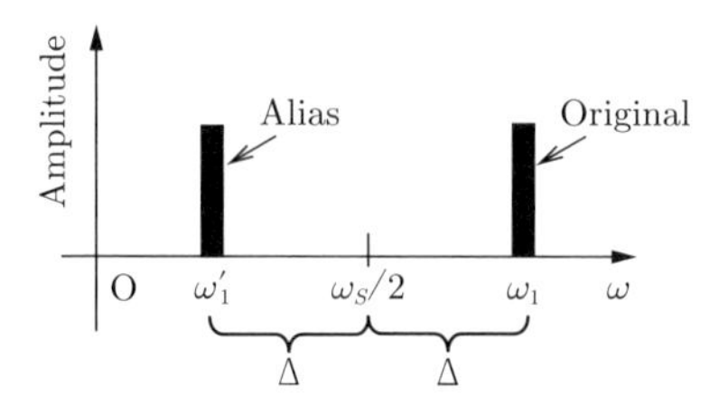

Figure 5.7 Aliasing

(5.3) shows that angular-frequency components higher than the Nyquist angular frequency, ω_N $(= \omega_S/2)$, do not precisely sampled. Thus, we only need to show the spectrum of a signal up to ω_N.

5.2 Selection of Sampling Period

The lowest sampling angular frequency should be at least twice the highest angular frequency component in a signal, ω_M. In practice, we usually choose a sampling angular frequency much higher than $2\omega_M$.

Considering the relationship between an angular frequency ω and a time period T is $T = 2\pi/\omega$, we discuss this issue in the time domain in this section.

† fold back ～：～を折り返す（畳む・曲げる）。

Let the period of a component be T_P seconds. A reasonable choice of a sampling period ΔT is

$$\frac{T_P}{\Delta T} \geqq 10, \tag{5.4}$$

that is, the sampling period is selected less than or equal to one-tenth of the period of a component.

We use an example to show the influence of the sampling period for analysis. The Great Hanshin Earthquake (Kobe earthquake) occurred on January 17, 1995, in Hyogo Prefecture, Japan. The moment magnitude scale was 6.9 and the maximum intensity was 7 [the JMA (Japan Meteorological Agency) seismic intensity scale]. The original ground acceleration of the Kobe wave and its spectrum are shown in **Figure 5.8**.

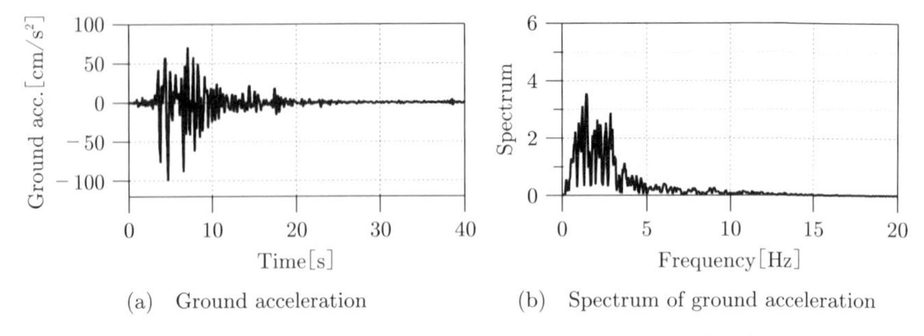

(a) Ground acceleration (b) Spectrum of ground acceleration

Figure 5.8 Original ground acceleration and spectrum of Kobe wave

It shows that the main components of the wave are in the frequency range $[1, 3]$ Hz, that is, $[0.33, 1]$ s. The period of the shortest main component is $T_P = 1/3 = 0.33$ s. We selected the sampling period to be $\Delta T = 0.033, 0.1, 0.33$, and 1 s. The sampled data and its spectrum are almost the same as the original for $T_P/\Delta T = 10$ (**Figure 5.9**). The main features of the wave are reserved for $T_P/\Delta T = 3$, which satisfies the requirement given by the sampling theorem (**Figure 5.10**). However, the frequency components are lost if their frequencies are higher than 1.5 Hz

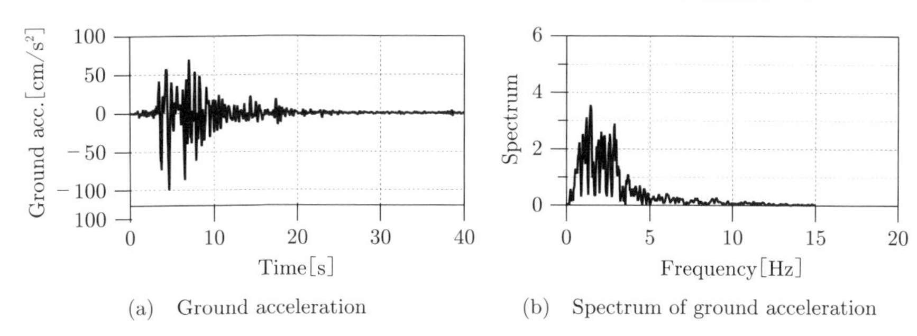

(a) Ground acceleration (b) Spectrum of ground acceleration

Figure 5.9 Ground acceleration and spectrum of Kobe wave for $\Delta T = 0.033\,\mathrm{s}$

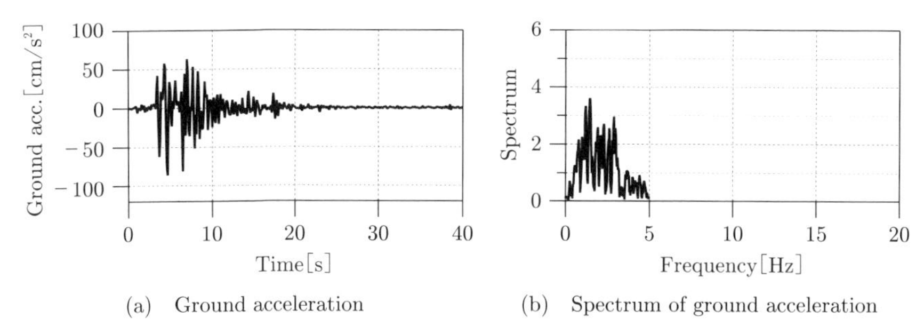

(a) Ground acceleration (b) Spectrum of ground acceleration

Figure 5.10 Ground acceleration and spectrum of Kobe wave for $\Delta T = 0.1\,\mathrm{s}$

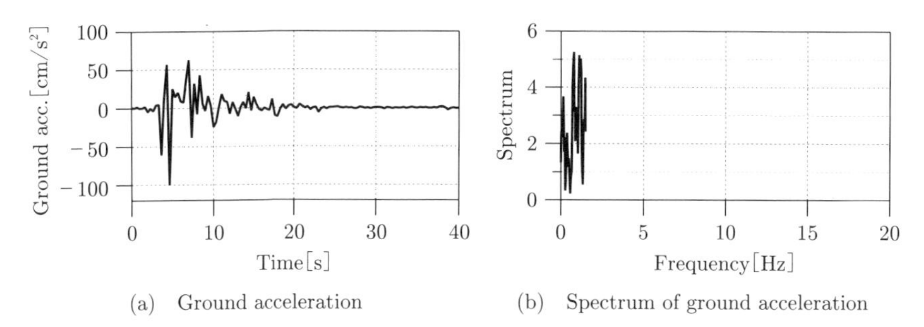

(a) Ground acceleration (b) Spectrum of ground acceleration

Figure 5.11 Ground acceleration and spectrum of Kobe wave for $\Delta T = 0.33\,\mathrm{s}$

for $T_P/\Delta T = 1$ (**Figure 5.11**) and if their frequencies are higher than $0.5\,\mathrm{Hz}$ for $T_P/\Delta T = 0.33$ (**Figure 5.12**). As a result, the shapes of the sampled waves are quite different from the original one. To ensure precise

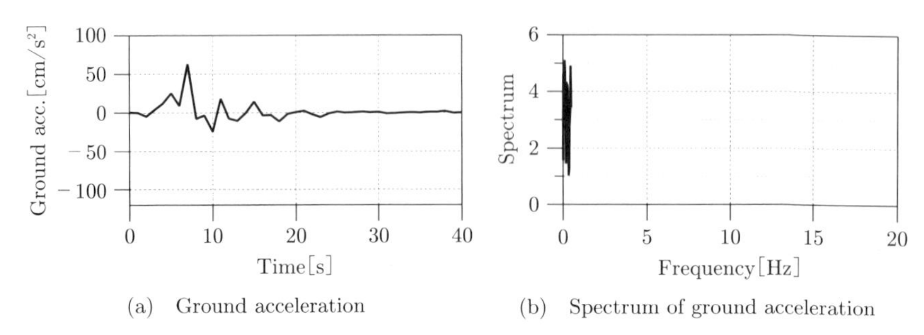

(a) Ground acceleration (b) Spectrum of ground acceleration

Figure 5.12 Ground acceleration and spectrum of Kobe wave for $\Delta T = 1\,\mathrm{s}$

analysis of a seismic wave, a sampling period is usually chosen to be $0.01\,\mathrm{s}$ in structural engineering.

5.3 Reconstruction of Signal from Its Samples

We can use different methods to reconstruct a continuous signal from its samples. However, it is important to note that reconstruction is not the inverse of sampling and only produces one possible continuous-time signal that has the same samples as a given discrete time signal.

We explain two methods for this purpose: a *zero-order hold* (ゼロ次ホールド) and a *first-order hold* (一次ホールド). A zero-order hold is a rectangular function

$$h_{\mathrm{ZOH}}(t) = \begin{cases} 1, & \text{if } 0 \leqq t < \Delta T, \\ 0, & \text{otherwise}, \end{cases} \tag{5.5}$$

where ΔT is a sampling period. For a discrete sequence $\{s[k]\}$ $(k = 0, 1, 2, \ldots)$, if we use a zero-order hold to reconstruct a continuous signal, it yields a *piecewise-constant* (区分的に一定) signal†

† We use $s(t)$ to describe a continuous signal and $s[i]$ to describe a discrete sequence.

$$s_{\mathrm{ZOH}}(k\Delta T+\tau)=s[k],\quad 0\leqq\tau<\Delta T,\quad k=0,1,2,\ldots,\tag{5.6}$$

or

$$s_{\mathrm{ZOH}}(t)=\sum_{k=0}^{\infty}s[k]h_{\mathrm{ZOH}}(t-k\Delta T),\tag{5.7}$$

that is, the value $s[k]$ is held during $[k\Delta T,(k+1)\Delta T)$.

A first-order hold is a triangular function described by

$$h_{\mathrm{FOH}}(t)=\begin{cases}1-\dfrac{|t|}{\Delta T}, & \text{if } |t|<\Delta T,\\ 0, & \text{otherwise.}\end{cases}\tag{5.8}$$

A piecewise-linear continuous signal is given by

$$s_{\mathrm{FOH}}(k\Delta T+\tau)=s[k]+\frac{s[k]-s[k-1]}{\Delta T}\tau,$$
$$0\leqq\tau<\Delta T,\quad k=0,1,2,\ldots,\tag{5.9}$$

or

$$s_{\mathrm{FOH}}(t)=\sum_{k=0}^{\infty}s[k]h_{\mathrm{FOH}}\left(t-k\Delta T\right).\tag{5.10}$$

Note that $s[k]=0$ for $k<0$.

The results of reconstructing a sine signal using zero-order and first-order holds are shown in **Figure 5.13**.

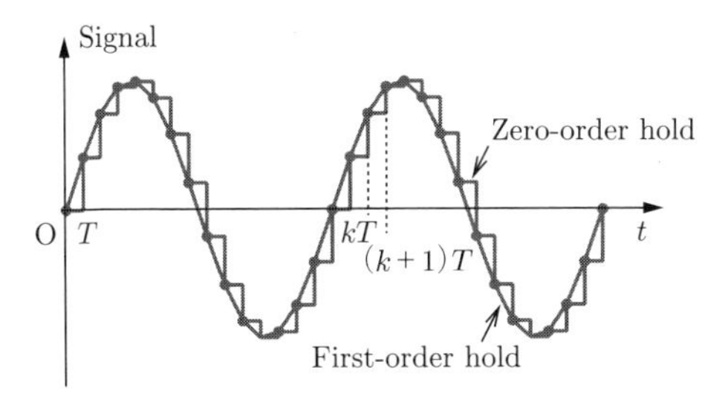

Figure 5.13 Reconstructing signal using zero- and first-order holds

Problems

Basic Level

【1】 Explain the effect if a sampling period is not small enough.

【2】 Consider the problem of sampling a signal $s(t) = \cos 10\pi t + \sin 20\pi t + \cos 30\pi t$. Find the condition for a sampling period at which aliasing occurs.

【3】 The sampling theorem specifies the lowest sampling frequency required to recover the highest frequency in a signal. Illustrate the difference between the reconstructed signals for an ideal sampling timing and others.

【4】 A signal with a solid line in **Figure 5.14** contains an oscillation component of π rad/s. If we sample it with a sampling period of 2 s ($\Delta T = 2$ s), we obtain a sampled signal without oscillation; and if we sample it with $\Delta T = 1.8$ s, we obtain a sampled signal with oscillation at a lower frequency. Explain these phenomena and calculate the oscillation angular frequency for $\Delta T = 1.8$ s.

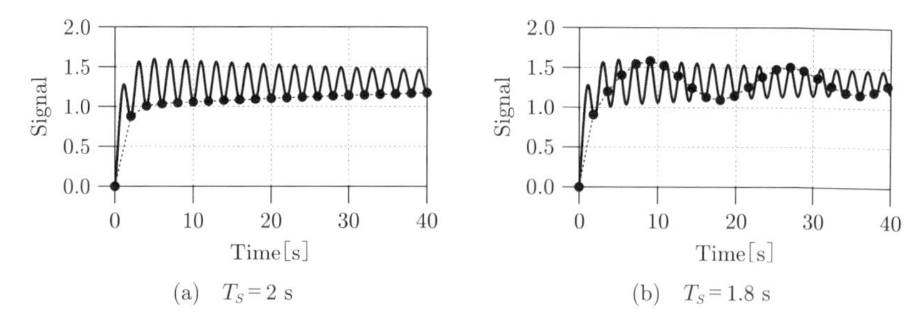

Figure 5.14 Original (solid) and sampled (dotted) signals with sampling period being 2 s and 1.8 s

【5】 The human hearing range is roughly [20, 20000] Hz. Explain why the sampling frequency $f = 44.1$ kHz is for audio signals.

【6】 Use the zero-order hold in Simulink to reconstruct $s(t) = \sin 2\pi t + \sin 6\pi t + \sin 10\pi t$ (**Figure 5.15**). Set different sampling periods in the hold and verify its effect.

Advanced Level

【1】 The commercial power supply in Tokyo is 100 V, 50 Hz. The period of the

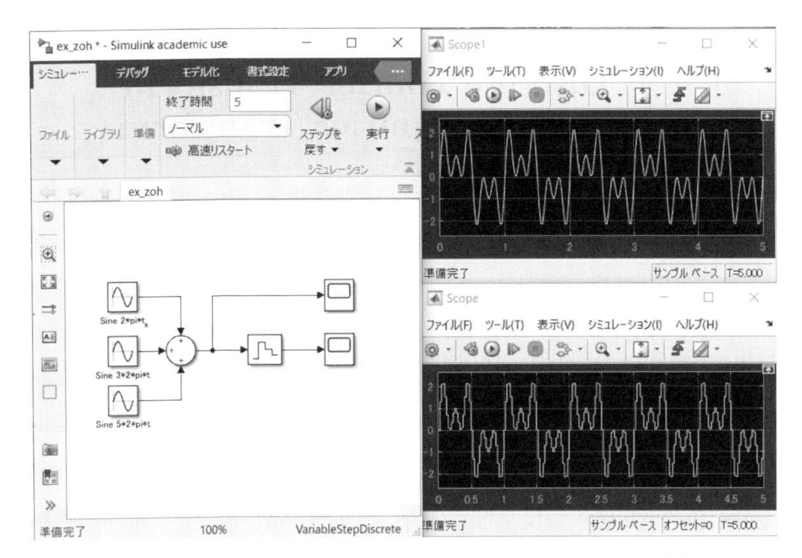

Figure 5.15 Signal reconstruction using zero-order hold

commercial voltage is 0.02 (= 1/50) s. Adding the voltage [**Figure 5.16**(a)] to a full-wave rectification yields the waveform in Figure 5.16(b). The period of the output voltage, $s(t)$, is 0.01 s. If a sampling period for the output is chosen to be 0.01, 0.01/2, 0.01/10, 0.01/256, and 0.01/1024 s, use the commands `fft` and `ifft` in MATLAB to calculate the spectra of the output for those choices and compare the results for different sampling periods (Carefully examine the sample program `05_ad_prob1.m`).

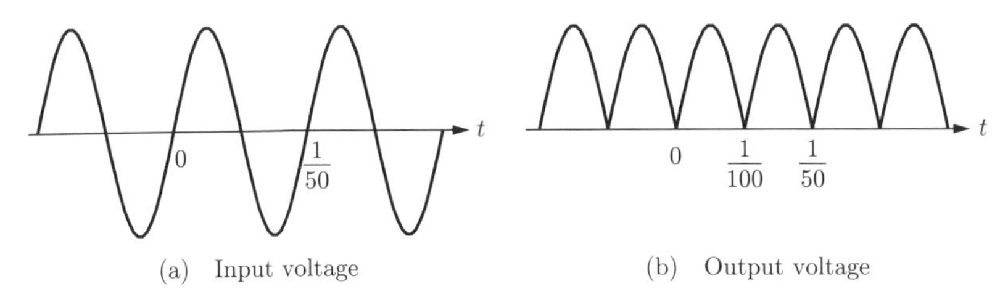

(a) Input voltage (b) Output voltage

Figure 5.16 Input and output voltages of full-wave rectification

[2] Differentiation does not affect the bandwidth of a signal: If the bandwidth of a band-limited signal $s(t)$ is ω_M, determine the Nyquist angular frequency for the signal $\mathrm{d}s(t)/\mathrm{d}t$.

6 Discrete Fourier Transform and Fast Fourier Transform

When we analyze a signal, we usually handle digital data collected from experiments. Unlike a continuous signal, a digital one is limited both in time and frequency bands. As explained in Chapter 5, if a high enough sampling frequency is used to collect samples from a continuous signal, a discrete version of the signal is a reasonable approximation of the original continuous one. This allows us to carry out the Fourier analysis on a discrete signal to find its frequency characteristics.

In 1965, Cooley and Tukey devised an algorithm to perform the Fourier transform for a discrete signal[6]. The algorithm greatly reduced the computational load and accelerated the computational speed. It gave the basic idea of a *fast Fourier transform* (FFT, 高速フーリエ変換) and accelerated it to be widely used in many fields since then.

6.1 Discrete Fourier Transform

The Fourier transform of a signal $f(t)$ is given by

$$F(\omega) = \int_{-\infty}^{\infty} f(t)\mathrm{e}^{-j\omega t}\mathrm{d}t, \tag{6.1}$$

and is indicated by $F(\omega) = \mathcal{F}[f(t)]$.

The sampling theorem establishes a fundamental bridge between continuous-time signals (that is, analog signals) and discrete-time signals (that is, digital signals). It presents a sufficient condition for a sampling rate to ensure that a continuous-time signal can be represented in its samples and can be recovered back.

Discrete a continuous-time signal $f(t)$ using a sampling period ΔT yields $f_k = f(k\Delta T)$ $(k = 0, 1, 2, \ldots, N-1)$. Then,

$$\begin{aligned} F(\omega) &= \int_0^{N\Delta T} f(t)\mathrm{e}^{-j\omega t}\mathrm{d}t \\ &\approx f_0\Delta T + f_1\mathrm{e}^{-j\omega\Delta T}\Delta T + \cdots + f_k\mathrm{e}^{-j\omega k\Delta T}\Delta T + \cdots \\ &\quad + f_{N-1}\mathrm{e}^{-j\omega(N-1)\Delta T}\Delta T \\ &= \Delta T \sum_{k=0}^{N-1} f_k \mathrm{e}^{-j\omega k\Delta T}. \end{aligned} \tag{6.2}$$

While we can evaluate the signal for any ω in principle, we only have N data points. So, only N final outputs are significant.

A continuous Fourier series evaluates a signal over its fundamental period. Similarly, we treat f_k as if it were periodic, that is, $\{f_N, f_{N+1}, \ldots, f_{2N-1}\}$ is the same as $\{f_0, f_1, \ldots, f_{N-1}\}$.

Note that the lowest angular frequency is

$$\omega_0 = \frac{2\pi}{N\Delta T}. \tag{6.3}$$

Choosing

$$\omega_n = n\omega_0 = \frac{2\pi n}{N\Delta T}, \quad n = 0, 1, \ldots, N-1 \tag{6.4}$$

yields

$$F(\omega_n) = \Delta T \sum_{k=0}^{N-1} f_k \mathrm{e}^{-jn\omega_0 k\Delta T} = \Delta T \sum_{k=0}^{N-1} f_k \mathrm{e}^{-j\frac{2\pi nk}{N}}. \tag{6.5}$$

On the other hand, from the definition of the inverse Fourier transform, we have

$$f(t) = \frac{1}{2\pi}\int_{-\infty}^{\infty} F(\omega)\mathrm{e}^{j\omega t}\mathrm{d}\omega \approx \frac{1}{2\pi}\sum_{n=0}^{N-1} F(\omega_n)\mathrm{e}^{jn\omega_0 t}\omega_0. \tag{6.6}$$

Thus, choosing $t = k\Delta T$ $(k = 0, 1, 2, \ldots, N-1)$ provides us with

$$f_k = \frac{1}{2\pi} \sum_{n=0}^{N-1} \left(\Delta T \sum_{m=0}^{N-1} f_m e^{-j\frac{2\pi nm}{N}} \right) e^{j\frac{2\pi n}{\Delta TN} k\Delta T} \frac{2\pi}{\Delta TN}$$
$$= \frac{1}{N} \sum_{n=0}^{N-1} \left(\sum_{m=0}^{N-1} f_m e^{-j\frac{2\pi nm}{N}} \right) e^{j\frac{2\pi kn}{N}}. \tag{6.7}$$

As a result, we define the *discrete Fourier transform* (DFT, 離散フーリエ変換) of a sampled signal $f_k = f(k\Delta T)$ $(k = 0, 1, 2, \ldots, N-1)$ to be

$$F_n = \sum_{m=0}^{N-1} f_m e^{-j\frac{2\pi nm}{N}}, \ n = 0, 1, \ldots, N-1 \tag{6.8}$$

and the *inverse discrete Fourier transform* (IDFT, 離散フーリエ逆変換) to be

$$f_k = \frac{1}{N} \sum_{n=0}^{N-1} F_n e^{j\frac{2\pi kn}{N}}, \quad k = 0, 1, 2, \ldots. \tag{6.9}$$

Comparing (6.8) and (6.5) shows that there is a ΔT-fold relationship between a Fourier transform and its discrete Fourier transform.

Let

$$W_N^1 = e^{-j\frac{2\pi}{N}}, \tag{6.10}$$

which is called a rotator. Multiplying it rotates data by $-2\pi/N$ degrees (see **Figure 6.1** for $N = 8$). Applying (2.10) yields $W_N^N = 1$ and $W_N^{kN+r} = W_N^r$ $(k = 0, 1, 2, \ldots, 0 \leqq r < N)$. Using W_N^1 to rewrite (6.8) gives

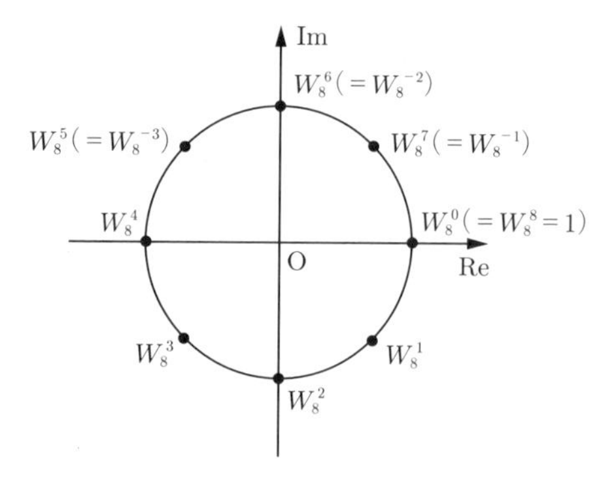

Figure 6.1 Rotator for $N = 8$

$$F_n = \sum_{m=0}^{N-1} f_m W_N^{mn}, \; n = 0, 1, \ldots, N-1, \tag{6.11}$$

which equals

$$\begin{cases} F_0 = f_0 + f_1 + f_2 + \cdots + f_{N-1} \\ F_1 = f_0 + f_1 W_N^1 + f_2 W_N^2 + \cdots + f_{N-1} W_N^{N-1} \\ F_2 = f_0 + f_1 W_N^2 + f_2 W_N^4 + \cdots + f_{N-1} W_N^{N-2} \\ \quad \vdots \\ F_{N-1} = f_0 + f_1 W_N^{N-1} + f_2 W_N^{N-2} + \cdots + f_{N-1} W_N^1. \end{cases} \tag{6.12}$$

Writing (6.12) in a matrix form yields

$$\begin{bmatrix} F_0 \\ F_1 \\ F_2 \\ \vdots \\ F_{N-1} \end{bmatrix} = \begin{bmatrix} 1 & 1 & 1 & \cdots & 1 \\ 1 & W_N^1 & W_N^2 & \cdots & W_N^{N-1} \\ 1 & W_N^2 & W_N^4 & \cdots & W_N^{N-2} \\ \vdots & \vdots & \vdots & \vdots & \vdots \\ 1 & W_N^{N-1} & W_N^{N-2} & \cdots & W_N^1 \end{bmatrix} \begin{bmatrix} f_0 \\ f_1 \\ f_2 \\ \vdots \\ f_{N-1} \end{bmatrix}. \tag{6.13}$$

【Example 6.1】 Calculate the DFT of $\{f_0, f_1, f_2, f_3\} = \{1, 1, -1, -1\}$.

Since $N = 4$, we have $W_4^1 = \mathrm{e}^{-j\frac{2\pi}{4}} = -j$, $W_4^2 = -1$, and $W_4^3 = j$. Thus,

$$\begin{bmatrix} F_0 \\ F_1 \\ F_2 \\ F_3 \end{bmatrix} = \begin{bmatrix} 1 & 1 & 1 & 1 \\ 1 & -j & -1 & j \\ 1 & -1 & 1 & -1 \\ 1 & j & -1 & -j \end{bmatrix} \begin{bmatrix} 1 \\ 1 \\ -1 \\ -1 \end{bmatrix} = \begin{bmatrix} 0 \\ 2-j2 \\ 0 \\ 2+j2 \end{bmatrix}. \tag{6.14}$$

On the other hand, $\omega_0 = 2\pi/(4\Delta T) = \pi/(2\Delta T)$, $2\omega_0 = \pi/\Delta T$, and $3\omega_0 = 3\pi/(2\Delta T)$. The amplitude of each frequency is $|F_0| = 0$ for $\omega = 0$, $|F_1| = 2\sqrt{2}$ for $\omega = \omega_0$, $|F_2| = 0$ for $\omega = 2\omega_0$, and $|F_3| = 2\sqrt{2}$ for $\omega = 3\omega_0$.

Note that there are $(N-1)$ times of addition for F_0, and $(N-1)$ times of multiplication and $(N-1)$ times of addition for each of $F_1, F_2, \ldots, F_{N-1}$ in (6.12). Thus, $(N-1)^2 \approx N^2$ (for a large N) times of multiplication and $N(N-1) \approx N^2$ times of addition in total are required to calculate F_n $(n = 0, 1, \ldots, N-1)$. Considering that the calculation of multiplication takes much longer time than that of addition, we usually count the number of multiplication to estimate computational complexity. Clearly, the complexity of a DFT has the order of N^2, which is indicated by $O(N^2)$. It is computationally expense when N is large.

6.2 Fast Fourier Transform

When the sample number of a signal increases, the computational complexity of its DFT increases explosively. Thus, calculating the DFT of a signal with a large number of samples is often very slow and is not practical.

An FFT was derived based on elegantly using the periodicity of $W_N^{\pm k\pi}$. It reduces the computational complexity to $O(N \log_2 N)$. For example, if $N = 1000$, then $N^2 = 10^6$ and $N \log_2 N \approx 10^4$. The computational load of the FFT is only about 1% of the corresponding DFT. The FFT provides us with an important practical tool for signal analysis. Gilbert Strang described an FFT as "the most important numerical algorithm of our lifetime"†.

An FFT is carried out using a *bit-reversal permutation* (ビット反転順列). We explain it as follows: First, consider a DFT for $N = 2$:

$$\begin{bmatrix} F_0 \\ F_1 \end{bmatrix} = \begin{bmatrix} 1 & 1 \\ 1 & W_2^1 \end{bmatrix} \begin{bmatrix} f_0 \\ f_1 \end{bmatrix}. \tag{6.15}$$

† G. Strang: Wavelets, American Scientist, 82 3, pp.250–255 (1994)

Note that $W_2^1 = \mathrm{e}^{-j\frac{2\pi}{2}} = -1$,

$$\begin{cases} F_0 = f_0 + f_1, \\ F_1 = f_0 - f_1. \end{cases} \tag{6.16}$$

Describing (6.16) in a figure yields a 2-point FFT (**Figure 6.2**).

Figure 6.2 FFT for $N = 2$

Now, consider a DFT for $N = 4$:

$$\begin{bmatrix} F_0 \\ F_1 \\ F_2 \\ F_3 \end{bmatrix} = \begin{bmatrix} 1 & 1 & 1 & 1 \\ 1 & -j & -1 & j \\ 1 & -1 & 1 & -1 \\ 1 & j & -1 & -j \end{bmatrix} \begin{bmatrix} f_0 \\ f_1 \\ f_2 \\ f_3 \end{bmatrix}. \tag{6.17}$$

Swapping the second and third raws gives its FFT:

$$\begin{bmatrix} F_0 \\ F_2 \\ F_1 \\ F_3 \end{bmatrix} = \begin{bmatrix} \begin{pmatrix} 1 & 1 \\ 1 & -1 \end{pmatrix} & \begin{pmatrix} 1 & 1 \\ 1 & -1 \end{pmatrix} \\ \begin{pmatrix} 1 & -j \\ 1 & j \end{pmatrix} & -\begin{pmatrix} 1 & -j \\ 1 & j \end{pmatrix} \end{bmatrix} \begin{bmatrix} f_0 \\ f_1 \\ f_2 \\ f_3 \end{bmatrix}. \tag{6.18}$$

Clearly, we only need two transformation matrices

$$E_1 = \begin{bmatrix} 1 & 1 \\ 1 & -1 \end{bmatrix}, \ E_2 = \begin{bmatrix} 1 & -j \\ 1 & j \end{bmatrix} \tag{6.19}$$

to carry out its FFT.

Now, we state the above discussion in a systematic way. The DFT for $N = 4$ is

$$\begin{bmatrix} F_0 \\ F_1 \\ F_2 \\ F_3 \end{bmatrix} = \begin{bmatrix} 1 & 1 & 1 & 1 \\ 1 & W_4^1 & W_4^2 & W_4^3 \\ 1 & W_4^2 & W_4^4 & W_4^2 \\ 1 & W_4^3 & W_4^2 & W_4^1 \end{bmatrix} \begin{bmatrix} f_0 \\ f_1 \\ f_2 \\ f_3 \end{bmatrix}. \tag{6.20}$$

Substituting $W_4^1 = -j$ into (6.20), rearranging the equations of F_i ($i = 0, 1, 2, 3$) in the order of bit reversal [represent the same binary bits of i ($i = 0, 1, 2, 3$) in the reversed order] [(F_0, F_2, F_1, F_3) in **Table 6.1**], and dividing them into two groups of equal size yield

$$\begin{cases} \begin{cases} F_0 = (f_0 + f_2) + (f_1 + f_3), \\ F_2 = (f_0 + f_2) - (f_1 + f_3), \end{cases} \\ \begin{cases} F_1 = (f_0 - f_2) + W_4^1(f_1 + W_4^2 f_3) = W_4^0(f_0 - f_2) + W_4^1(f_1 - f_3), \\ F_3 = (f_0 - f_2) + W_4^1(W_4^2 f_1 + f_3) = W_4^0(f_0 - f_2) - W_4^1(f_1 - f_3). \end{cases} \end{cases} \tag{6.21}$$

Table 6.1 2-bit reversal

Original		Bit reversal	
Decimal	Binary	Binary	Decimal
0 (F_0)	00	00	0 (F_0)
1 (F_1)	01	10	2 (F_2)
2 (F_2)	10	01	1 (F_1)
3 (F_3)	11	11	3 (F_3)

Thus, the top two and the bottom two equations in (6.21) can be computed using the 2-point FFT (**Figure 6.3**).

In the same manner, an FFT for $N = 8$ is shown in **Figure 6.4**. Note that $W_8^0 = W_4^0$ and $W_8^2 = W_4^1$, and the 3-bit reversal is shown in **Table 6.2**.

Now, considering the general case of an FFT. Let the number of data points be a power of 2, that it, $N = 2^n$ ($n \in \mathbb{N}_0$†). It has three steps:

† $\mathbb{N}_0$ is the set of non-negative integers.

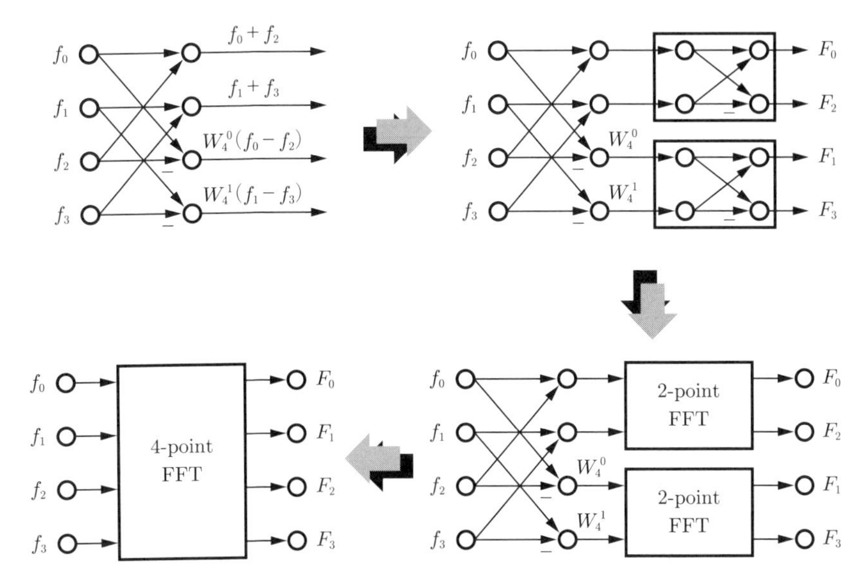

Figure 6.3 FFT for $N = 4$

Step 1: Arrange f_i $(i = 0, 1, \ldots, N-1)$ in time order. Then, divide them into two parts: the first and the rest $N/2$-tuple data. Carry out the calculation shown in (a) in **Figure 6.5**.

Step 2: Send the processed first and the rest $N/2$-tuple data to the $N/2$-point FFT.

Step 3: Rearrange F_n to $F_0, F_1, \ldots, F_{N-1}$ based on the n-bit reversal and output the result.

A great number of FFT algorithms have been developed to make an FFT fast and efficient to match the requirements of various applications. FFT algorithms are widely used in speech, image, data processing, and other fields, and are available in many numerical computational packages, such as MATLAB and Scilab, as well as exclusive packages†.

† For example, https://www.kurims.kyoto-u.ac.jp/˜ooura/fft-j.html, http://www2.kobe-u.ac.jp/˜kuroki/OpenSoft/FFT2D/index.html.

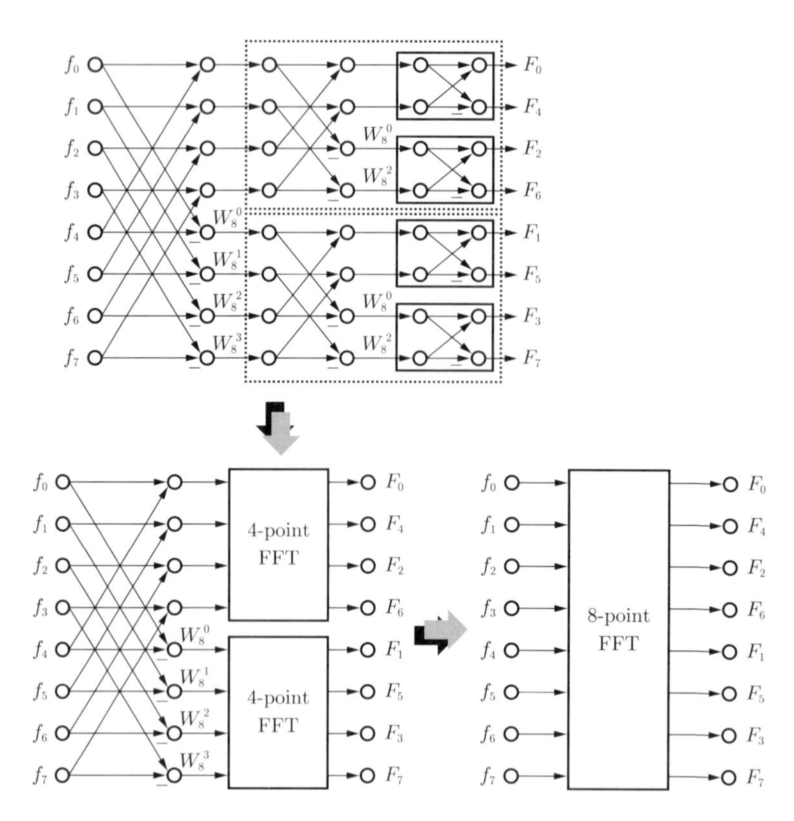

Figure 6.4 FFT for $N = 8$

Table 6.2 3-bit reversal

Original		Bit reversal	
Decimal	Binary	Binary	Decimal
0 (F_0)	000	000	0 (F_0)
1 (F_1)	001	100	4 (F_4)
2 (F_2)	010	010	2 (F_2)
3 (F_3)	011	110	6 (F_6)
4 (F_4)	100	001	1 (F_1)
5 (F_5)	101	101	5 (F_5)
6 (F_6)	110	011	3 (F_3)
7 (F_7)	111	111	7 (F_7)

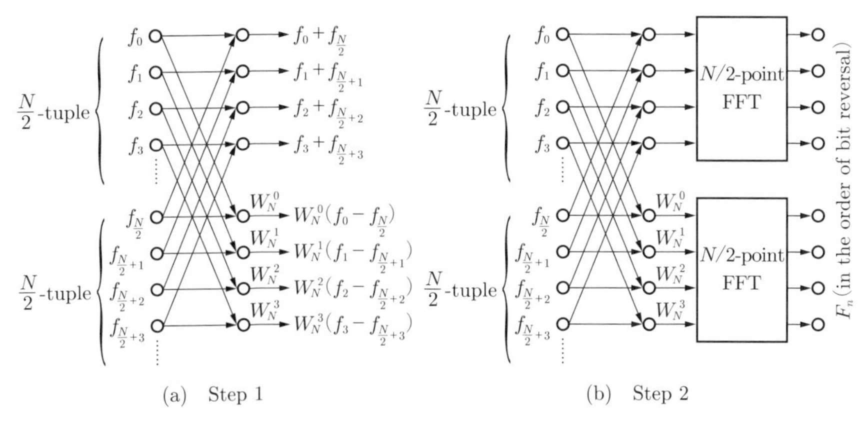

Figure 6.5 FFT algorithm

【Example 6.2】 Use MATLAB to calculate the power spectrum of

$$s(t) = \sin 2\pi 50t + \sin\left(2\pi 70t + \frac{\pi}{4}\right) + n(t), \tag{6.22}$$

where $n(t)$ is a white noise.

First, choosing the sampling time to be $\Delta T = 0.001\,\mathrm{s}$, we produce a time sequence of the signal (6.22). Then, we use the command `fft` in MATLAB to find the FFT of the signal. Since the sampling angular frequency is $\omega_S = 2\pi/\Delta T = 2000\pi\,\mathrm{rad/s}$, The Nyquist angular frequency is $\omega_N = \omega_S/2 = 1000\pi\,\mathrm{rad/s}$. So, we only need to calculate the spectrum up to ω_N.

The discrete version of Parseval's identity in Subsection 4.2 is

$$\sum_{k=0}^{N-1} |f_k|^2 = \frac{1}{N} \sum_{n=0}^{L-1} |F_n|^2. \tag{6.23}$$

The MATLAB commands for the example are listed in Program 6.1. Note that `randn` generates a white noise, `std` calculates the standard deviation of the signal, `db2mag` converts decibel to magnitude, and `std(x)/db2mag(5)` designates the signal-to-noise ratio to 5 dB.

Program 6.1 FFFexmp

```
% Calculate the power spectrum of a signal

% Build a noised signal sampled at 1000 Hz containing pure
    frequencies at 50 and 70 Hz
dT=0.001; % sampling period: 0.001 s
fs=1/dT; % sampling rate: 1000 Hz
N=500; % number of sample points
t = (0:N-1)/fs;
x=sin(2*pi*50*t)+sin(2*pi*70*t+pi/4)+randn(size(t));
    % signal
s = x + randn(size(x)).*std(x)/db2mag(5); % S/N = 5 dB

clf()
figure(1)
plot(t,s); grid on;

% FFT of signal s
n=length(s); % original sample number
L=pow2(nextpow2(n)); % length for FFT

y=fft(s, L); % FFT of signal s
power1=abs(y).^2/L; % power of signal s

fN=(0:L/2)*(fs/L); % frequency range
power=power1(1:L/2+1);

figure(2)
plot(fN,power); grid on;
xlabel('Frequency [Hz]'); ylabel('Power spectrum');
```

Problems

Basic Level

【1】 Use the definition to calculate the DFTs of the following signals:

A. $f_k = 1\ (k = 0, 1, 2, \ldots, N-1)$.

$$\text{B.}\quad f_k = \begin{cases} 1, & \text{if } k = 0, \\ 0, & \text{if } k = 1, 2, \ldots, N-2, \\ -1, & \text{if } k = N-1. \end{cases} \tag{6.24}$$

【2】 Consider the DFT of a periodic signal f_k with the period N ($f_{N+k} = f_k$). Prove that $F_{N+k} = F_k$.

【3】 Calculate the 4-point DFTs of the signals in **Figure 6.6**.

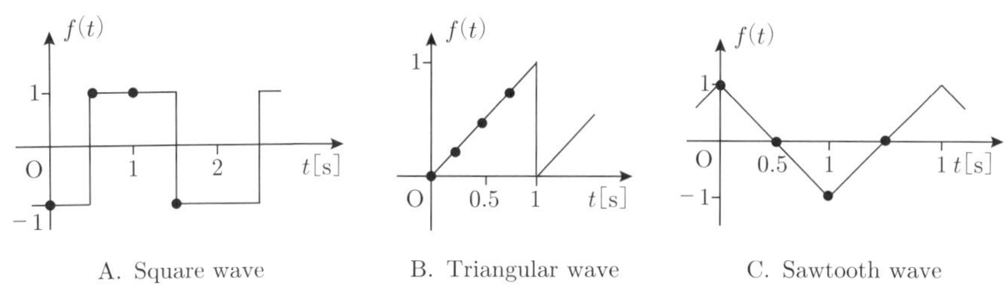

Figure 6.6 Basic-Level Problem 3

【4】 Calculate the 8-point DFTs of the signals in **Figure 6.7**.

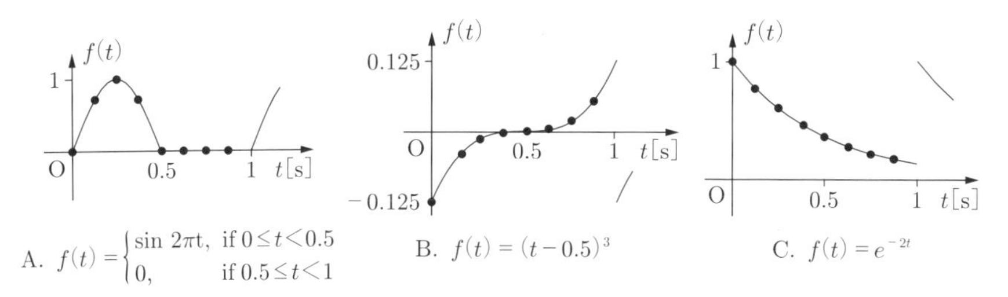

Figure 6.7 Basic-Level Problem 4

【5】 Calculate the 4-point FFTs of the signals in Basic-Level Problem 3.

【6】 Calculate the 8-point FFTs of the signals in Basic-Level Problem 4.

【7】 Refer to Example 6.2 and use the command `fft` in MALTAB to calculate the power spectrum of $f(t) = \cos 4\pi t + n(t)$ [$n(t)$: noise] with the sampling period of:

A. 0.01 s. B. 0.1 s. C. 0.25 s. D. 0.5 s. E. 1 s.

And discuss the calculated results.

【8】 Use the command `fft` in MALTAB to calculate the power spectrum of $f(t) = 1 + \cos \pi t + \sin \pi t + n(t)$ [$n(t)$: noise]. Examine the results for the sampling period of:

A. 2 s. B. 1 s. C. 0.5 s. D. 0.25 s. E. 0.01 s. F. 0.001 s.

And discuss how the sampling period should be chosen.

Advanced Level

【1】 Prove Parseval's identity (6.23).

【2】 Answer the following questions for $f(t) = \sin 2\pi t + \sin 4\pi t + \sin 6\pi t$:

A. Calculate the Fourier transform of $f(t)$, $F(\omega)$.

B. Use the command `fft` in MALTAB to calculate the 8-point FFT of $f(t)$, F_n $(n = 0, 1, \ldots, 7)$.

C. Plot the real parts $\mathrm{Re}[F(\omega)]$ and $\mathrm{Re}(F_n)$ $(n = 0, 1, \ldots, 7)$.

D. Plot the Imaginary parts $\mathrm{Im}[F(\omega)]$ and $\mathrm{Im}(F_n)$ $(n = 0, 1, \ldots, 7)$.

E. Verify that the Fourier coefficient c_0 is the average of the signal.

【3】 Explain that the computational load of an FFT is $O(N \log_2 N)$, where N is the number of sample data.

【4】 A window function is a function that is zero-valued outside of a prescribed interval. When a window function is multiplied over a signal, the values outside the interval become zero. It leaves only the values inside the finite interval. Thus, it eliminates the need for an infinite number of calculations and facilitates numerical analysis. For this reason, it is widely used in signal processing and statistics.

Sample a sine signal $f(t) = \sin 100\pi t$ at a sampling rate of 1 kHz. A length-L portion of $f(t)$ is windowed by a rectangular and a Hamming window, that is,

$$\begin{cases} f_k^{\mathrm{rec}} = w_{\mathrm{rec}} \times f_k = 1 \times \sin 0.1\pi k, \\ f_k^{\mathrm{ham}} = w_{\mathrm{ham}} \times f_k = \left(0.54 - 0.46 \cos \dfrac{2\pi k}{L-1}\right) \sin 0.1\pi k. \end{cases} \tag{6.25}$$

Choose L to be 100 and 200. Calculate the power spectra of f_k^{rec} and f_k^{ham}.

7 Applications to Engineering Problems

A Fourier transform is an important technique for the analysis of sounds, solid-state properties (based on X-rays, ultraviolet rays, infrared rays, etc.), and images in many industrial fields.

This chapter explains two well-known engineering applications of the Fourier transform: sound and vibration analysis.

7.1 Analysis of Sound

A sound is a wave, that is, its amplitude changes with time. *Phonetics* (音声学) studies how sounds are produced and perceived. It has three categories: articulatory, acoustic, and auditory. *Articulatory phonetics* (調音音声学) physiologically analyzes how humans produce sounds through the interaction of different physiological structures. *Acoustic phonetics* (音響音声学) physically analyzes acoustic aspects of sounds. And *auditory phonetics* (聴覚音声学) psychologically concerns the hearing and interpreting of sounds.

Acoustic phonetics studies the physical properties of sounds from four aspects: intensity, loudness, wavelength, and timbre. A sound wave is usually decomposed into different frequency components using the FFT to perform analysis.

Acoustic features of a sound include its frequency (pitch) and amplitude a spectrum and a *spectrogram* (スペクトログラム), *formant* (形成音, フォルマン

ト, ホルマント), *cepstrum* (ケプストラム), *Mel-frequency cepstral coefficients* (メル周波数ケプストラム係数), and linear predictive coefficients.

A sound is measured in terms of frequency and amplitude. The loudness of a sound is measured by its amplitude. The fundamental frequency and its harmonics, which are the integer multiples of the fundamental frequency, are important parameters of a sound (Harmonics are also called natural resonances.).

A spectrum is a sound pressure or power as a function of frequency. It shows the intensity level for each frequency that makes up the sound. Let $F(\omega)$ be the Fourier transform of a continuous-time sound $f(t)$. The energy of the sound is

$$E = \int_{-\infty}^{\infty} |f(t)|^2 \, \mathrm{d}t. \tag{7.1}$$

Parseval's identity provides us with

$$E = \frac{1}{2\pi} \int_{-\infty}^{\infty} |F(\omega)|^2 \, \mathrm{d}\omega. \tag{7.2}$$

The energy spectral density (ESD) of the sound is

$$\mathrm{ESD}(\omega) = |F(\omega)|^2 . \tag{7.3}$$

The definitions of ESD and E are suitable for a transient signal, such as a pulse-like signal, because their energy is finite and we can find their Fourier transforms.

For a continuous signal over all time, let

$$\hat{f}(t) = \begin{cases} f(t), & \text{if } -\dfrac{T}{2} \leqq t \leqq \dfrac{T}{2}, \\ 0, & \text{otherwise} \end{cases} \tag{7.4}$$

and

$$F(\omega) = \int_{-\infty}^{\infty} \hat{f}(t) \mathrm{e}^{-j\omega t} \mathrm{d}t = \int_{-T/2}^{T/2} f(t) \mathrm{e}^{-j\omega t} \mathrm{d}t. \tag{7.5}$$

Since E given in (7.1) and (7.2) usually becomes infinity as T tends to infinity, we define the average power of $f(t)$ to be

$$P = \lim_{T\to\infty} \frac{1}{T} \int_{-T/2}^{T/2} |f(t)|^2 \, \mathrm{d}t \tag{7.6}$$

and the *power spectral density* (PSD, パワースペクトル密度) to be

$$\mathrm{PSD}(\omega) = \lim_{T\to\infty} \frac{1}{T} |F(\omega)|^2 = \lim_{T\to\infty} \frac{1}{T} \mathrm{ESD}(\omega). \tag{7.7}$$

PSD describes how the power of a signal is distributed over frequency.

Note that $\mathrm{PSD}(-\omega) = \mathrm{PSD}(\omega)$ holds if $f(t)$ is real. In some applications, we only consider the integration range $[0, \infty)$ in (7.5) and (7.7). A PSD defined over such a range is called a single-sided PSD. If we use $\mathrm{PSD1}(\omega)$ to indicate it and $\mathrm{PSD2}(\omega)$ to indicate a two-sided PSD [integrated over the range $(-\infty, \infty)$], then we have the following relationship based on (7.2):

$$\int_{-\infty}^{\infty} \mathrm{PSD2}(\omega) \mathrm{d}\omega = \int_{0}^{\infty} \mathrm{PSD1}(\omega) \mathrm{d}\omega. \tag{7.8}$$

Thus,

$$\mathrm{PSD1}(\omega) = 2 \times \mathrm{PSD2}(\omega). \tag{7.9}$$

On the other hand, since it is impossible to observe a signal over a time span $[0, \infty)$ in experiments, we choose a suitable length T and define the average power and the power spectral density to be

$$P = \frac{1}{T} \int_{-T/2}^{T/2} |f(t)|^2 \, \mathrm{d}t \tag{7.10}$$

and the power spectral density to be

$$\mathrm{PSD}(\omega) = \frac{1}{T} |F(\omega)|^2 \tag{7.11}$$

in practice.

For a discrete-time sound, f_k, let its discrete Fourier transform be F_n. Then, its energy spectral density is

$$\mathrm{ESD}_N = |F_n|^2, \tag{7.12}$$

and its power spectral density is

$$\mathrm{PSD}_N = \frac{1}{N}|F_n|^2, \tag{7.13}$$

where N is the number of samples of f_k.

A spectrogram is a result of calculating the frequency spectrum of a composite signal with a window function [it is called a *short-time Fourier transform* (短時間フーリエ変換)]. It shows the result of signal frequency spectrum analysis with time on the horizontal axis, frequency on the vertical axis, and signal strength expressed in colors and shades.

Let $w(t)$ be a window function that is zero-valued outside a chosen interval. Then, $w(t-t_0)$ is the function that shifts $w(t)$ by t_0 on the time axis. Thus, the short-time Fourier transform of $f(t)$ is

$$F(\omega, t) = \int_{-\infty}^{\infty} f(\tau)w(\tau - t)\mathrm{e}^{-j\omega t}\mathrm{d}\tau. \tag{7.14}$$

Let the length of a window be T. Some widely used window functions (**Figure 7.1**) are the rectangular window

$$w(t) = \begin{cases} 1, & \text{if } 0 \leqq t \leqq T, \\ 0, & \text{otherwise,} \end{cases} \tag{7.15}$$

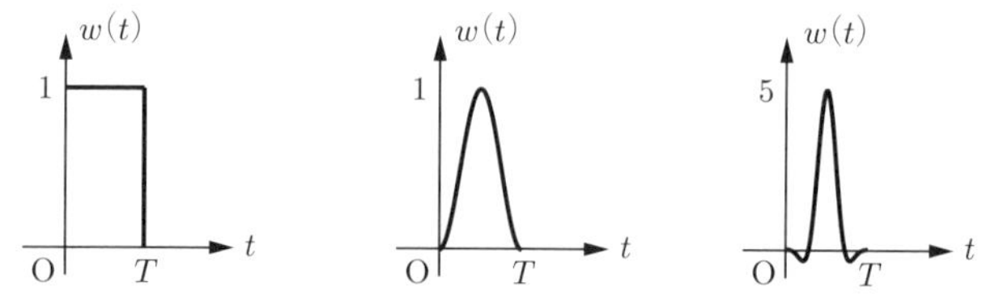

(a) Rectangular window (b) Hanning window (c) Flat-top window

Figure 7.1 Window functions

the Hanning window

$$w(t) = \begin{cases} 0.5(1 - \cos 2\pi t), & \text{if } 0 \leqq t \leqq T, \\ 0, & \text{otherwise,} \end{cases} \tag{7.16}$$

and the flat-top window

$$w(t) = \begin{cases} 1 - 1.93\cos 2\pi t + 1.29\cos 4\pi t & \\ \quad - 0.388\cos 6\pi t + 0.032\cos 8\pi t, & \text{if } 0 \leqq t \leqq T, \\ 0, & \text{otherwise.} \end{cases} \tag{7.17}$$

Table 7.1 shows the use of the window functions.

Table 7.1 Window functions

Name	Frequency resolution	Application
Rectangular	Good	Transient signals such as impulse waveforms
Hanning	Fair	General continuous signals (most common)
Flat-top	Poor	Amplitude-accurate-measurement-oriented signals such as high-frequency analysis

If we let ΔT be a sampling time, $f_k = f(k\Delta T)$, and $w_{k-m} = w(k\Delta T - m\Delta T)$, then the discrete-time version of the short-time Fourier transform for f_k $(k = 0, \ldots, N-1)$ is

$$F(n, m) = \sum_{k=0}^{N-1} f_k w_{k-m} \mathrm{e}^{-j\frac{2\pi kn}{N}}. \tag{7.18}$$

Human hearing has the characteristic of being less sensitive to changes in pitch (sound height) for higher tones. A scale that takes this perceptual characteristic into account is the *mel scale* (メルスケール). It is a scale of pitches that human hearing perceives as equidistant. When the frequency of a spectrum is converted to the mel scale, the spectrum is called a *mel spectrum* (メルスペクトル). A common formula to convert hertz into mel is (**Figure 7.2**)

$$M = 2595\log_{10}\left(1 + \frac{f}{700}\right). \tag{7.19}$$

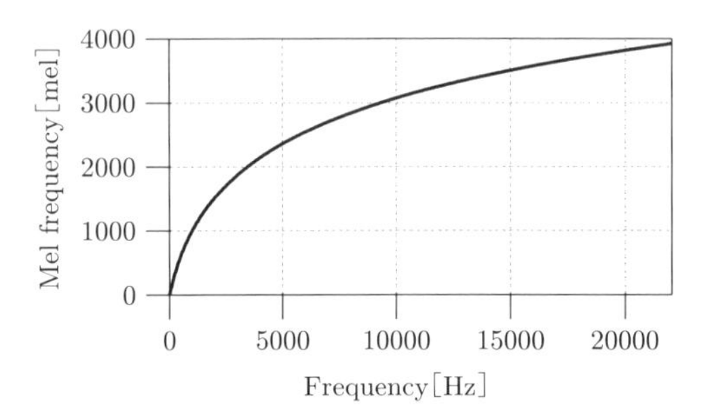

Figure 7.2 Mel versus Hz

When a sound signal contains turbulence, the power spectrum of a source signal changes more smoothly than that of turbulence. Cepstrum analysis uses this characteristic to separate these two signals. Except for the analysis of sound and vibration, it has also been used in the analysis of an electrocardiogram (ECG) signal, a heart sound, and other biomedical signals.

A *cepstrum* (ケプストラム) was originally defined to be a power cepstrum, which is the inverse Fourier transform of a power spectrum. For a general signal, $f(t)$, we can define a real cepstrum and a complex cepstrum as follows:

$$\begin{cases} \text{Real cepstrum: } C_r = \mathcal{F}^{-1}\left\{\log_{10}|\mathcal{F}[f(t)]|\right\} \\ \text{Complex cepstrum: } C_c = \mathcal{F}^{-1}\left\{\log_{10}\mathcal{F}[f(t)]\right\}. \end{cases} \tag{7.20}$$

Note that

$$C_c = \mathcal{F}^{-1}\left\{\log_{10}\left[|F(\omega)|e^{j\angle F(\omega)}\right]\right\} = \mathcal{F}^{-1}\left\{\log_{10}|F(\omega)| + j\frac{\angle F(\omega)}{\log_{10} e}\right\}. \tag{7.21}$$

If we take the logarithm of a power cepstrum at each mel-frequency, carry out a cosine transform of the logarithmic function, and collect the amplitudes of mel-frequency cepstra, then *mel-frequency cepstral coefficients* (MFCCs, メル周波数ケプストラム係数) are coefficients that collectively make up a mel-frequency cepstrum.

【Example 7.1】 Crows in Tokyo and other large cities in Japan cause various troubles. In 2001, the Tokyo Metropolitan Government established a crow management project team and launched a comprehensive campaign. The efforts have produced certain achievements, On the other hand, this problem has been tackled from an engineering standpoint[†1]. A strategy for inducing escape behavior in crows was successfully devised based on the analysis of crow calls. It plays the calls representing alarm, threat, and escape from a loudspeaker in sequence with appropriate timing to control the behavior of crows.

In the following example, we analyze some warning calls of a kind of crow called Corvus macrorhynchos.

First, download audio files (`alarm.wav` and `threat.wav`) from the site `https://crowlab.co.jp/service/index.html#sounds`[†2] and save them in the current directory. Next, considering that the calls are not stationary, we split the original sound data [**Figure 7.3**(a)] into five frames (`alarm1.wav, alarm2.wav, ..., alarm5.wav`) [Figure 7.3(b)] based on the cries. Then, we analyze each of the frames. Taking `alarm1.wav` as an example, we list the MATLAB commands in Program 7.1.

After using the Hanning window for the sound (**Figure 7.4**), we calculate the spectrum and the PSD of the signal in both the linear and logarithmic scales (**Figure 7.5**). Then, we calculate the complex cepstrum (**Figure 7.6**) and the MFCCs (**Figure 7.7** shows the result of the fifth MFCC, MFCC 5.).

---

†1 https://crowlab.co.jp

†2 Courtesy of Dr. Naoki Tsukahara. http://tsukaharanaoki.net/

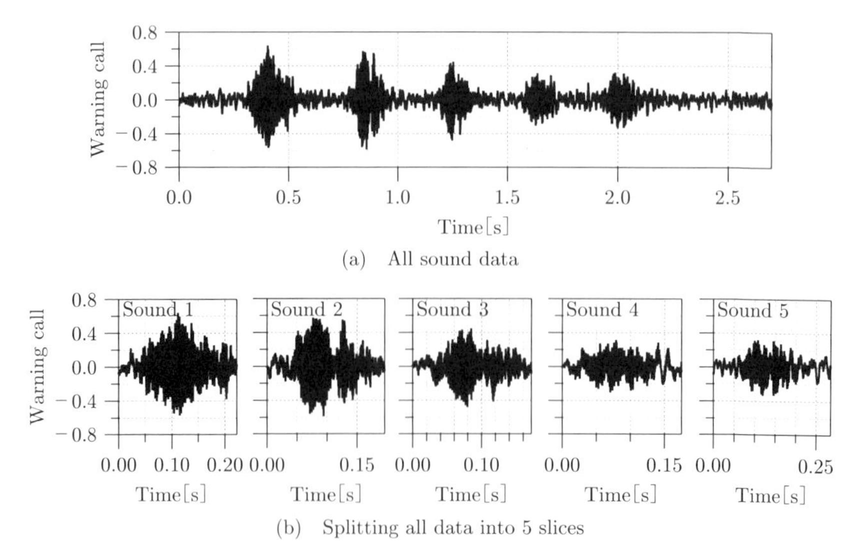

(a) All sound data

(b) Splitting all data into 5 slices

Figure 7.3 Crow's warning call

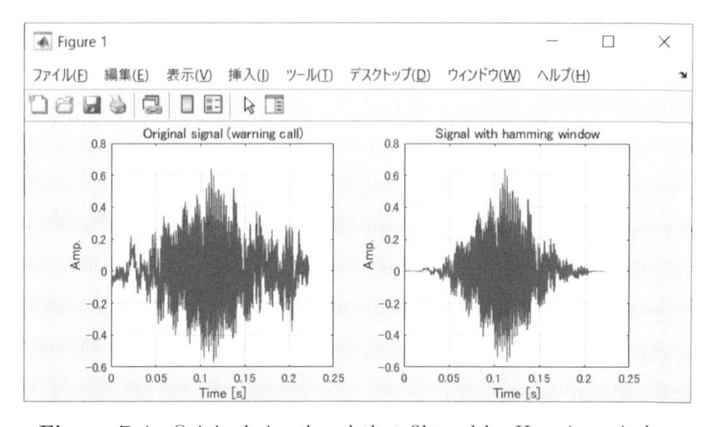

Figure 7.4 Original signal and that filtered by Hanning window

Program 7.1 FFFsound

```
1  % Sound processing program
2
3  fname = fullfile('alarm1.wav'); % A slice of alarm.wav
4
5  [s0,Fs] = audioread(fname);
```

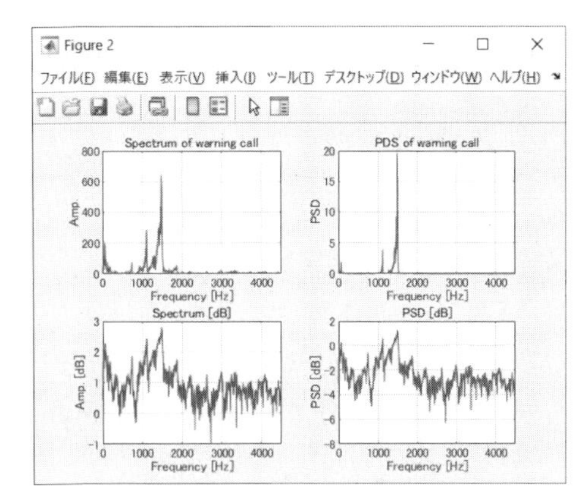

Figure 7.5 Spectrum and PSD

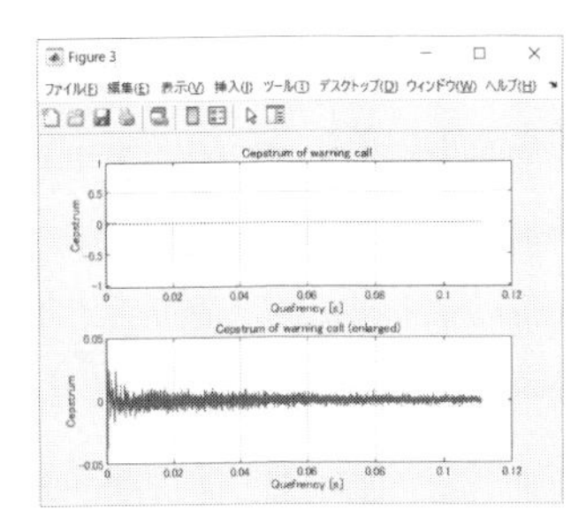

Figure 7.6 Cepstrum

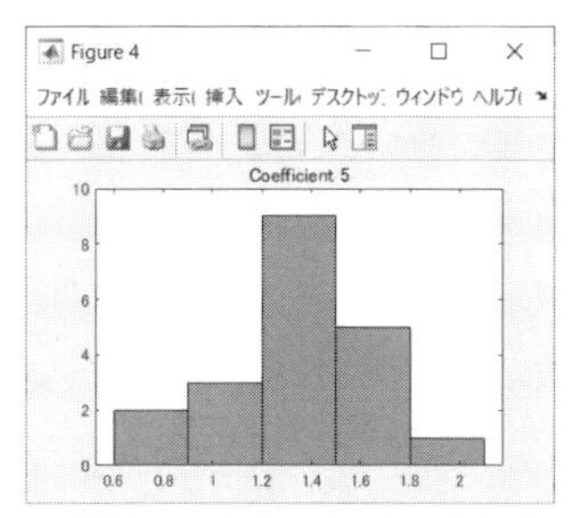

Figure 7.7 MFCC 5

```
6  sound(s0,Fs); % Play the sound
7  info = audioinfo(fname); % Info about audio file
8
9  t0 = 0:1/Fs:info.Duration;
10 t0 = t0(1:end-1); % Time duration
11
12 ns0=size(s0);
13 L1=ns0(1);
14 if rem(L1,2) == 1
15     L=L1-1; s=s0(1:L); t=t0(1:L); % Set L to be even
16 else
17     L=L1; s=s0; t=t0;
```

```
18 end
19
20 w=hann(L); % Hanning window
21 y=s.*w;
22
23 % Signal plots
24 figure(1)
25 subplot(1,2,1);
26 plot(t,s), grid
27 title('Original signal (warning call)');
28 xlabel('Time [s]'); ylabel('Amp.');
29
30 subplot(1,2,2);
31 plot(t,y), grid
32 title('Signal with Hamming window');
33 xlabel('Time [s]'); ylabel('Amp.');
34
35 % FFT of original sound
36 yFFT = fft(y);
37
38 yFFT_abs = abs(yFFT); % Magnitude of FFT
39 Afft_log2=log10(yFFT_abs); % Amp. spectrum [dB]
40 ESD2 = power(abs(yFFT),2); % Energy spectral density
41 PSD2 = ESD2/L; % PSD of sound
42
43 yFFT_abs1=yFFT_abs(1:L/2+1);
44 yFFT_abs1(2:end-1)=2*yFFT_abs1(2:end-1); % Single-sided
45 Afft_log1=log10(yFFT_abs1); % Single-sided [dB]
46
47 PSD1=PSD2(1:L/2+1);
48 PSD1(2:end-1)=2*PSD1(2:end-1); % Single-sided PSD [dB]
49
50 % FFT and PSD plots
51 f=Fs*(0:(L/2))/L;
```

```
52
53 figure(2);
54 subplot(2,2,1);
55 plot(f(1:1000),yFFT_abs1(1:1000)); grid
56 title('Spectrum of warning call');
57 xlabel('Frequency [Hz]'); ylabel('Amp.');
58
59 subplot(2,2,2);
60 plot(f(1:1000),PSD1(1:1000)); grid
61 title('PDS of warning call');
62 xlabel('Frequency [Hz]'); ylabel('PSD');
63
64 subplot(2,2,3);
65 plot(f(1:1000), Afft_log1(1:1000)); grid
66 title('Spectrum [dB]');
67 xlabel('Frequency [Hz]'); ylabel('Amp. [dB]');
68
69 subplot(2,2,4);
70 plot(f(1:1000), PSD_log1(1:1000)); grid
71 title('PSD [dB]');
72 xlabel('Frequency [Hz]'); ylabel('PSD [dB]');
73
74 % Cepstrum
75 c=real(ifft(Afft_log2));
76 quefrency=linspace(0, info.Duration, info.Duration*Fs);
77
78 figure(3)
79 subplot(2,1,1);
80 plot(quefrency(1:L/2),c(1:L/2)), grid
81 title('Cepstrum of warning call')
82 xlabel('Quefrency [s]'); ylabel('Cepstrum');
83 subplot(2,1,2);
84 plot(quefrency(1:L/2),c(1:L/2)), grid
85 title('Cepstrum of warning call (enlarged)')
```

```
86 xlabel('Quefrency [s]');
87 ylabel('Cepstrum'); ylim([-0.05,0.05])
88
89 % MFCCs
90 coeffs = mfcc(y,Fs,"LogEnergy","Ignore");
91 % Plot a probability density function for
92 % a mel-frequency cepstral coefficient
93 figure(4)
94 coefficientToAnalyze = 5; % Choose Coefficient 5 for
       display
95 histogram(coeffs(:,coefficientToAnalyze+1))
96 title(sprintf("Coefficient %d",coefficientToAnalyze))
```

7.2 Analysis of Seismic Wave

Fourier analysis is widely used in structural engineering to estimate earthquake and wind responses, perform active structural control, and so on.

This section explains how to use Fourier spectra to analyze seismic waves so that we can estimate the maximum response of a structure for a seismic wave.

Seismic isolation protects a building from seismic waves. The basic idea is to use a controller to adjust the fundamental frequency of a building to ensure that it is different from the predominant frequency of a seismic wave. Fourier analysis is used to obtain the energy spectrum and time-frequency distribution of an excitation input signal, derive a control algorithm, and verify a control result.

The motion of a structure is described by

$$M\ddot{x}(t) + C\dot{x}(t) + Kx(t) = f(t), \tag{7.22}$$

where M, C, and K are the mass, damping coefficient, and stiffness of the structure, respectively; $x(t)$ is the displacement of the structure; and $f(t)$ is an exogenous force. Performing the Fourier transform of (7.22) gives

$$(j\omega)^2 MX(\omega) + j\omega CX(\omega) + KX(\omega) = F(\omega), \tag{7.23}$$

where $X(\omega)$ and $F(\omega)$ are the Fourier transform of $x(t)$ and $f(t)$, respectively. Let $G(\omega)$ be

$$\begin{cases} G(\omega) = \dfrac{X(\omega)}{F(\omega)} = \dfrac{1/M}{(j\omega)^2 + 2\xi_s\omega_s(j\omega) + \omega_s^2}, \\ \omega_s = \sqrt{\dfrac{K}{M}}, \quad \xi_s = \dfrac{C}{2\sqrt{MK}}, \end{cases} \tag{7.24}$$

where ω_s [rad/s] is called the *natural angular frequency* (固有角周波数, 固有角振動数, 固有円振動数) of the structure, ξ_s is called the *damping ratio* (減衰係数), and $G(\omega)$ is the *transfer function* (伝達関数). It shows the amplification gain from the input $f(t)$ to the output $x(t)$. Clearly, the gain strongly depends on the frequency of an input signal. The relationship between ω and the gain $|G(\omega)|$ is nonlinear. On the other hand, if we use a logarithm to describe it

$$\text{Gain}(\omega) = 20\log_{10}|G(\omega)| \text{ [dB]}, \tag{7.25}$$

the gain can be approximated by polygonal lines. This greatly eases the analysis of system characteristics. Thus, we use a Bode plot to make it easy to show the relationship between Gain(ω) and ω. An example is shown in **Figure 7.8** for $\omega_s = 1.57$ rad/s (Natural period is 4.0 s and ξ_s is 0.20).

As explained in Section 5.2, the main components of the Kobe wave are in the frequency range of 1–3 Hz (6–8 rad/s), this structure has a very low gain for the wave. For example, the gain at 6 rad/s is about −30 dB, that

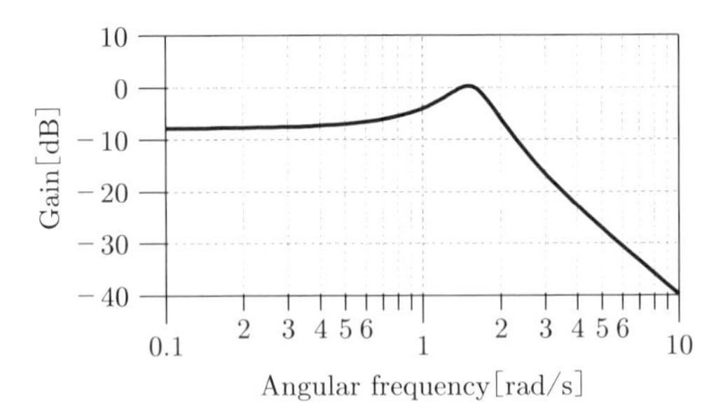

Figure 7.8 Frequency response of (7.24) for $\omega_s = 1.57\,\text{rad/s}$ and $\xi_s = 0.20$

is, the gain is $10^{-30/20} = 0.032$. Clearly, this wave gives little influence on this structure.

The convolution theorem in Section 4.2 reveals that the output of the structural response $x(t)$ for an exogenous force $f(t)$ is

$$x(t) = \int_0^\infty g(t-\tau)f(\tau)\mathrm{d}\tau, \tag{7.26}$$

$g(t)$ is the inverse Fourier transform of $G(\omega)$ in (7.24). It is given by

$$g(t) = \frac{1}{M\omega_d}\mathrm{e}^{-\xi_s\omega_s t}\sin\omega_d t, \quad \omega_d = \omega_s\sqrt{1-\xi_s^2}. \tag{7.27}$$

Substituting (7.27) into (7.26) yields

$$x(t) = \frac{1}{M\omega_d}\int_{-\infty}^\infty \mathrm{e}^{-\xi_s\omega_s(t-\tau)}\sin\omega_d(t-\tau)\times f(\tau)\mathrm{d}\tau. \tag{7.28}$$

Since $\ddot{x}_g(t) = 0$ for $t < 0$, $f(t) = 0$ holds for $t < 0$. As a result, (7.28) becomes

$$x(t) = \frac{1}{M\omega_d}\int_0^\infty \mathrm{e}^{-\xi_s\omega_s(t-\tau)}\sin\omega_d(t-\tau)\times f(\tau)\mathrm{d}\tau. \tag{7.29}$$

If $\xi_s = 0$, that is, the structure is undamped (the worst case), then we have

$$x(t) = \frac{1}{M\omega_s}\int_0^\infty \sin\omega_s(t-\tau)\times f(\tau)\mathrm{d}\tau. \tag{7.30}$$

The exogenous force $f(t)$ is

$$f(t) = M\ddot{x}_g(t), \tag{7.31}$$

where $\ddot{x}_g(t)$ is the ground acceleration. Substituting (7.31) into (7.30) gives

$$x(t) = \frac{1}{\omega_s}\int_0^{\infty} \sin\omega_s(t-\tau) \times \ddot{x}_g(\tau)\mathrm{d}\tau. \tag{7.32}$$

Let the Fourier transform of the ground motion $a_g(t)[= \ddot{x}_g(t)]$ be $A_g(\omega)$ and $G_a(\omega) = MG(\omega)$. It is clear that $G_a(\omega) = X(\omega)/A_g(\omega)$. It follows from convolution in the time domain, (4.23), that the frequency-domain version of (7.32) is

$$X(\omega) = \frac{1}{(j\omega)^2 + \omega_s^2}A_g(\omega) = G_a(\omega)|_{\xi_s=0}A_g(\omega). \tag{7.33}$$

The velocity of the structure is $v(t) = \mathrm{d}x(t)/\mathrm{d}t$. The derivative property given in Section 4.2 provides us with

$$V(\omega) = j\omega X(\omega) = [j\omega G(\omega)][(j\omega)^2 X_g(\omega)]. \tag{7.34}$$

Note that $\mathcal{F}^{-1}[j\omega G(\omega)] = \mathrm{d}g(t)/\mathrm{d}t$. The velocity of the structure for $\xi_s = 0$ is

$$\begin{aligned}
v(t) &= \frac{\mathrm{d}x(t)}{\mathrm{d}t} = \frac{1}{\omega_s}\int_0^{\infty} \frac{\mathrm{d}g(t-\tau)}{\mathrm{d}(t-\tau)} \times \ddot{x}_g(\tau)\mathrm{d}\tau \\
&= \frac{1}{M}\int_0^{\infty} \cos\omega_s(t-\tau) \times \ddot{x}_g(\tau)\mathrm{d}\tau \\
&= \frac{1}{M}\left[\int_0^{\infty} \ddot{x}_g(\tau)\cos\omega_s t\cos\omega_s\tau\mathrm{d}\tau + \int_0^{\infty} \ddot{x}_g(\tau)\sin\omega_s t\sin\omega_s\tau\mathrm{d}\tau\right] \\
&= \frac{1}{M}\left[\cos\omega_s t\int_0^{\infty} \ddot{x}_g(\tau)\cos\omega_s t\mathrm{d}\tau + \sin\omega_s t\int_0^{\infty} \ddot{x}_g(\tau)\sin\omega_s\tau\mathrm{d}\tau\right] \\
&= \bar{V}(\omega_s)\cos(\omega_s t + \phi_s),
\end{aligned} \tag{7.35}$$

where

$$\begin{cases} \bar{V}(\omega_s) = \dfrac{1}{M}\sqrt{\left[\displaystyle\int_0^\infty \ddot{x}_g(\tau)\cos\omega_s\tau \mathrm{d}\tau\right]^2 + \left[\displaystyle\int_0^\infty \ddot{x}_g(\tau)\sin\omega_s\tau \mathrm{d}\tau\right]^2}, \\ \phi_s = -\tan^{-1}\dfrac{\displaystyle\int_0^\infty \ddot{x}_g(\tau)\sin\omega_s\tau \mathrm{d}\tau}{\displaystyle\int_0^\infty \ddot{x}_g(\tau)\cos\omega_s\tau \mathrm{d}\tau}. \end{cases} \tag{7.36}$$

The maximum velocity $|v(t)|_{\max}$ is

$$|v(t)|_{\max} = \left|\bar{V}(\omega_s)\cos(\omega_s t + \phi_s)\right|_{\max} = |\bar{V}(\omega_s)|. \tag{7.37}$$

(7.32), (7.35), and (7.37) show how to calculate the time responses of the displacement, the velocity, and the maximum velocity of the structure to a ground motion, respectively. On the other hand, the frequency responses of the displacement and velocity are given in (7.33) and (7.34). Since the calculation of a frequency response is a simple arithmetic operation, it features easy to calculate. Note that the inverse Fourier transforms of $X(\omega)$ and $V(\omega)$ yield their time responses. (7.33) and (7.34) provide us with an easy way for this purpose. The frequency at which $X(\omega)$ or $V(\omega)$ has a peak is called the *dominant frequency* (卓越振動数). It means that the structure tends to shake at the frequency.

Furthermore, the Fourier transform of the ground motion is

$$\begin{aligned} A_g(\omega) &= \int_0^\infty \ddot{x}_g(t)\mathrm{e}^{-j\omega t}\mathrm{d}t \\ &= \int_0^\infty \ddot{x}_g(t)\cos\omega t\mathrm{d}t - j\int_0^\infty \ddot{x}_g(t)\sin\omega t\mathrm{d}t. \end{aligned} \tag{7.38}$$

Hence,

$$\begin{aligned} |A_g(\omega)| &= \sqrt{\left[\int_0^\infty \ddot{x}_g(t)\cos\omega\tau\mathrm{d}t\right]^2 + \left[\int_0^\infty \ddot{x}_g(t)\sin\omega t\mathrm{d}t\right]^2} \\ &= \frac{\bar{V}(\omega)}{M}, \end{aligned} \tag{7.39}$$

where

$$|\bar{V}(\omega)| = \frac{1}{M}\sqrt{\left[\int_0^\infty \ddot{x}_g(t)\cos\omega t\mathrm{d}t\right]^2 + \left[\int_0^\infty \ddot{x}_g(t)\sin\omega t\mathrm{d}t\right]^2}. \tag{7.40}$$

7.3 Processing of Surface Electromyography (sEMG)

A *surface electromyography* (sEMG) (表面筋電図) signal captures the state of muscle activity and motor function (**Figure 7.9**). The changes in sEMG signals correlate to the number of motor units, activity patterns, metabolic situations, and other factors. An† sEMG signal is nonstationary during muscle dynamic contractions and shows a high degree of complexity. Processing sEMG signals provides us with a way to obtain exact information about muscle status. It has widely been used in rehabilitation, sports science, and many other fields.

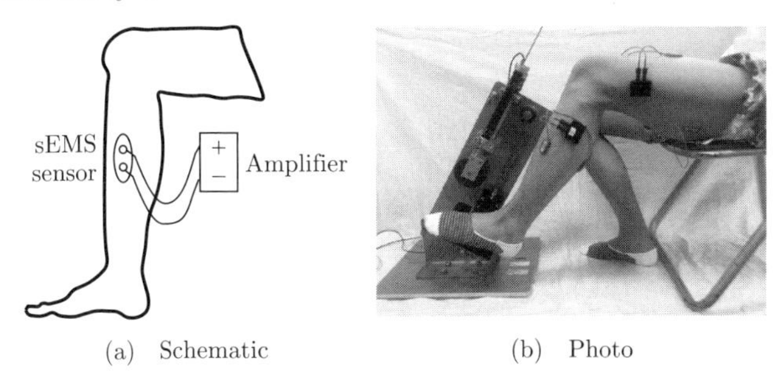

(a) Schematic (b) Photo

Figure 7.9 Measuring sEMG signal

Walking is a basic activity in daily life. *Loin muscle* (*psoas*) (腰筋), *gluteal muscle* (*gluteus maximus*) (大殿筋), *front thigh muscle* (*quadriceps*

† a and an: whether *a* or *an* should precede a noun depends on how the first syllable is pronounced: a is used if the first syllable begins with a consonant sound and an if it begins with a vowel sound (Handbook of Writing for the Mathematical Sciences, N. J. Higham, SIAM, 2020).

femoris) (大腿四頭筋), *hamstring muscle* (*biceps femoris*) (大腿二頭筋), *calf muscle* (*soleus*) (ヒラメ筋), *shin muscle* (*tibialis anterior*) (前脛骨筋), and some others are involved in walking (they are called walking muscles) (**Figure 7.10**). Thus, the rehabilitation of walking muscles ensures mental and physical soundness and is of great meaning. It is important to judge muscle fatigue to perform a suitable level of lower-limb rehabilitation. sEMG signals are used for this purpose and the Fourier transform is a basic tool for the analysis of sEMG signals. This section explains some useful indexes in the assessment of muscle activities.

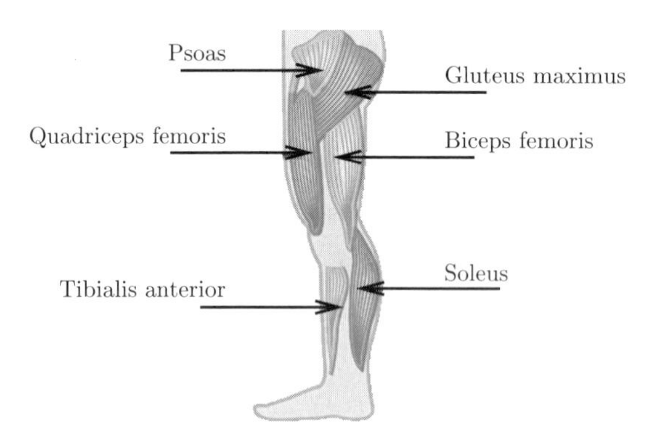

Figure 7.10 Walking muscles

A walking treadmill and an ergometer are widely used in lower-limb rehabilitation. We take pedaling an ergometer at a prescribed pedaling load as an example. Sampling the sEMG signal with the frequency of 1000 Hz obtains a signal $s(t)$ (**Figure 7.11**). The power spectrum of $s(t)$ is

$$P(f) = \frac{1}{T}|S(f)|^2, \tag{7.41}$$

where T and $S(f)$ are the duration time and the Fourier transform of $s(t)$, respectively [Note that $f = \omega/(2\pi)$].

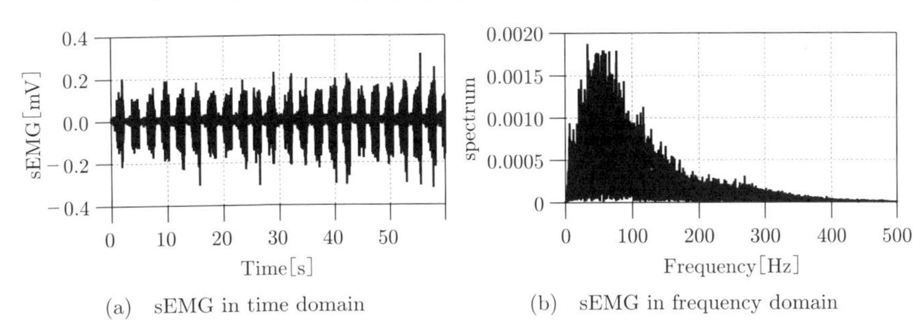

(a) sEMG in time domain (b) sEMG in frequency domain

Figure 7.11 sEMG signal of tibialis anterior and its Fourier transform

The frequencies of an sEMG signal are mainly in the range [20, 350] Hz. They are usually divided into three frequency bands: low ([20, 50) Hz), median ([50, 150) Hz), and high (≧ 150 Hz).

A median power frequency, f_c, is a frequency that satisfies

$$\int_0^{f_c} P(f)\mathrm{d}f = \int_{f_c}^{\infty} P(f)\mathrm{d}f \tag{7.42}$$

and a mean power frequency (MPF) is defined to be

$$\mathrm{MPF} = \frac{\displaystyle\int_0^{\infty} fP(f)\mathrm{d}f}{\displaystyle\int_0^{\infty} P(f)\mathrm{d}f}. \tag{7.43}$$

f_c and MPF are used as indexes of muscle fatigue. It is known that both f_c and MPF move to low frequencies as a muscle becomes fatigued.

Integrated sEMG (iEMG) is defined to be

$$\mathrm{iEMG} = \int_0^{T} |s(t)|\mathrm{d}t. \tag{7.44}$$

It is used to evaluate muscle activities and to determine muscle strength.

【Example 7.2】 In this example, we analyze the sEMGs of a front thigh muscle (`mus_data.csv`), which were collected during pedaling an ergometer for one minute at the sampling frequency of 1000 Hz.

The fundamental frequency of 50 Hz and its harmonic components generated by an AC power supply are artifacts in the sEMG signals. They were first removed using a digital IIR (infinite impulse response) filter (**Figure 7.12**). Note that the maximum frequency in a signal that can be reconstructed after sampling is the Nyquist frequency (Section 5.2) f_N ($= 500\,\mathrm{Hz}$). A normalized frequency $f_n = f/f_N$ is used in Figure 7.12, where f is the real sampling frequency and $f_n \in [0, 1]$.

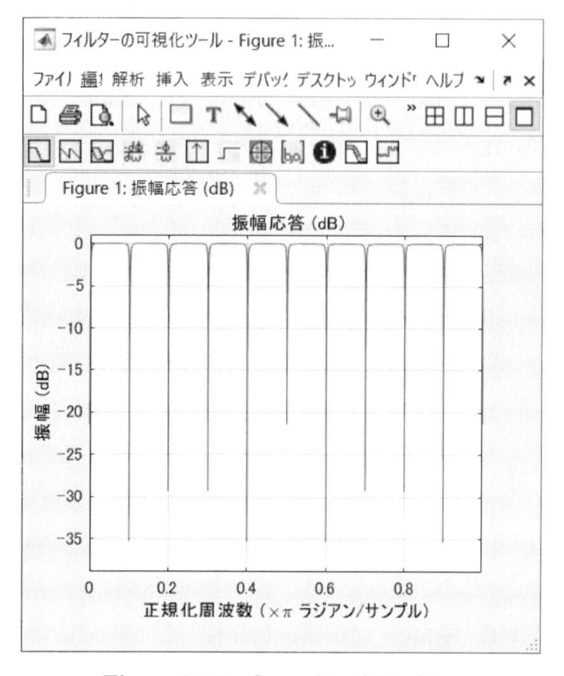

Figure 7.12 Gain plot of IIR filter

Signals with frequencies lower than 20 Hz are noise. Thus, we used a high-pass filter with a cutoff frequency of 20 Hz to remove the noise, and obtained filtered sEMG signals (**Figure 7.13**).

Then, we calculated the power spectral densities of the signals (**Figure 7.14**), the median frequencies (**Figure 7.15**)

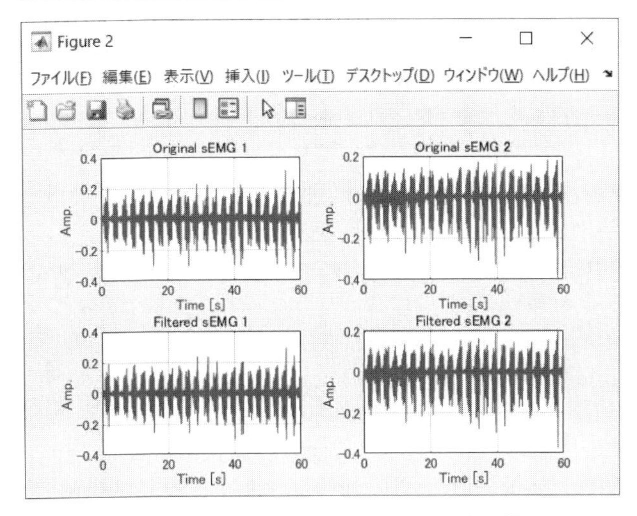

Figure 7.13 sEMG signals before and after filtering

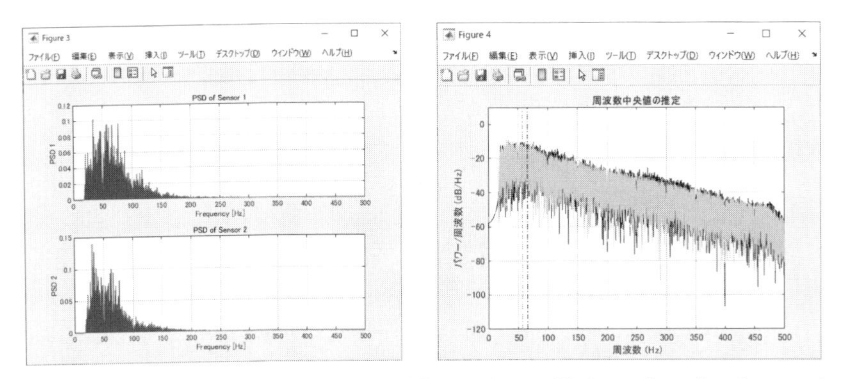

Figure 7.14 PSD of sEMG signals **Figure 7.15** Estimated median frequencies

$$f_{c1} = 65.88\,\mathrm{Hz},\ \ f_{c2} = 56.98\,\mathrm{Hz}, \tag{7.45}$$

and the integrated sEMGs

$$\mathrm{iEMG}_1 = 1.2239,\ \ \mathrm{iEMG}_1 = 1.2332. \tag{7.46}$$

Program 7.2 FFFsEMG

```
1 x0=load('mus_data.csv');
2 Fs=1000; % Sampling frequency
```

```
3  N=length(x0);
4  t0 =(0:N-1)./Fs; % Time interval [s]
5  FsN=Fs/2; % Nyquist frequency = 500 Hz
6
7  f0=50; % Notch frequency [Hz]
8  q=35;
9  bw=(f0/FsN)/q; % Bandwidth at the -3 dB point
10 [b,a] = iircomb(Fs/f0,bw,'notch'); % Note type flag
       'notch'
11 fvtool(b,a); % Notches falls at 50,100, ..., 500 Hz.
12 y0=filter(b,a,x0);
13
14 % highpass filter with cutoff frequency of 20 Hz
15 [y,d] = highpass(y0,20,Fs,'ImpulseResponse','iir','
       Steepness',0.5);
16
17 figure(1);
18 subplot(2,2,1);
19 plot(t0,x0(:,1)); grid
20 title('Original sEMG 1')
21 xlabel('Time [s]'); ylabel('Amp.');
22 subplot(2,2,2);
23 plot(t0,x0(:,2)); grid
24 title('Original sEMG 2')
25 xlabel('Time [s]'); ylabel('Amp.');
26
27 subplot(2,2,3);
28 plot(t0,y(:,1)); grid
29 title('Filtered sEMG 1')
30 xlabel('Time [s]'); ylabel('Amp.');
31 subplot(2,2,4);
32 plot(t0,y(:,2)); grid
33 title('Filtered sEMG 2')
34 xlabel('Time [s]'); ylabel('Amp.');
```

```
35
36 x1=y(:,1); x2=y(:,2);
37 L=length(x1);
38
39 % FFT
40 yf1=fft(x1); yf2=fft(x2);
41 yf1_abs=abs(yf1); yf2_abs=abs(yf2);
42 PSD2_1=power(abs(yf1),2)/L; PSD2_2=power(abs(yf2),2)/L;
43 PSD1_1=PSD2_1(1:L/2+1); PSD1_2=PSD2_2(1:L/2+1);
44 PSD1_1(2:end-1)=2*PSD1_1(2:end-1);
45 PSD1_2(2:end-1)=2*PSD1_2(2:end-1);
46
47 f = Fs*(0:(L/2))/L;
48 figure(3)
49 subplot(2,1,1);
50 plot(f,PSD1_1); grid
51 title('PSD of Sensor 1')
52 xlabel('Frequency [Hz]'); ylabel('PSD');
53 subplot(2,1,2);
54 plot(f,PSD1_2); grid
55 title('PSD of Sensor 2')
56 xlabel('Frequency [Hz]'); ylabel('PSD');
57
58 % Median frequency
59 medfreq([PSD1_1 PSD1_2],f)
60 fc1=medfreq(PSD1_1,f); % Sensor 1
61 fc2=medfreq(PSD1_2,f); % Sensor 2
62
63 % Integrated EMG
64 iEMG1=sum(abs(x1))/Fs
65 iEMG2=sum(abs(x2))/Fs
```

Problems

Basic Level

【1】 Understand MATLAB codes in Example 7.1 and use them to analyze other pieces of the sound `threat.wav` (`alarm2.wav-alarm5.wav`).

【2】 Use the Windows program called the voice recorder or other recording programs to record your own vocalizations of "a-i-u-e-o" and write your own MATLAB codes to carry out the following analysis:

A. Split the sound file into 5 pieces according to the vowels.

B. Using the rectangular, Hanning, and flat-top windows to operate on each piece.

C. Calculate the FFT, ESD, and PSD of each piece.

D. Calculate the real and complex cepstrums, and the MFCCs of each piece.

【3】 Download any audio data you like from AudioSet provided by Google (https://research.google.com/audioset/) and perform the analysis as those in Basic-Level Problem 2.

【4】 Download earthquake waves from the following sites

1) K-NET and Kik-NET
(https://www.kyoshin.bosai.go.jp/kyoshin/)

2) Japan Meteorological Agency
(https://www.data.jma.go.jp/eqev/data/kyoshin/jishin/index.html)

3) The Building Center of Japan
(https://www.bcj.or.jp/download/wave/).

Calculate and compare the dominant frequencies for earthquake waves.

【5】 Show that the average power per period for a periodic function $f_T(t)$ is $\frac{1}{T}\int_{-T/2}^{T/2} |f_T(t)|^2 \mathrm{d}t = \sum_{n=-\infty}^{\infty} |c_n|^2$, where T is the period of the function and c_n $(n = 0, \pm 1, \pm 2, \ldots)$ are the coefficients of the complex Fourier series of the function.

【6】 Understand MATLAB codes in Example 7.2 and use them to analyze sEMG signals downloaded from the following sites

1) IEEE (https://ieee-dataport.org/open-access/dataset-surface-electromyographic-semg-signals-and-finger-kinematics)

2) PhysioNet (https://physionet.org/content/hd-semg/1.0.0/)
3) putEMG (https://biolab.put.poznan.pl/putemg-dataset/)
4) CapgMyo (http://zju-capg.org/research_en_electro_capgmyo.html)
5) Ninaweb (http://ninaweb.hevs.ch/node/17).

Advanced Level

【1】 A speech signal is considered to be the composition of a source signal, such as vibrations of the vocal cords and turbulence caused by friction, and articulation determined by the shape of the vocal tract, oral cavity, and nasal cavity. A cepstrum is used to separate them. Conduct a literature review and explain it in detail.

【2】 The Mel spectrum and MFCCs are used as acoustic features that represent the outline of the spectrum while taking into account of human auditory characteristics. Conduct a literature review and explain them in detail.

【3】 Conduct a literature review and explain the use of the power spectral density, the median power frequency, the mean power frequency, and the integrated sEMG of an sEMG signal.

【4】 Let the natural frequency and the damping ratio of a structure be 3.14 rad/s and 0.02, respectively. Using the Fourier transform to calculate its response for a seismic wave downloaded from sources in Basic-Level Problem 6.

8 Application to Mathematical Problems in Engineering

Engineering mathematics is a branch of applied mathematics that applies mathematical methods and techniques to real-world problems in engineering and industry. Fourier analysis is one of the main tools used in this field. This chapter explains two main applications of Fourier analysis: One is impulse responses, which are widely used in system analysis; and the other is partial differential equations, which frequently appear in the mathematical analysis of various engineering problems.

8.1 Linear System and Impulse Response

A system is a mapping from an input $r(t)$ to an output $y(t)$ (**Figure 8.1**):

$$y(t) = \mathcal{T}[r(t)], \tag{8.1}$$

where $\mathcal{T}(\cdot)$ is a mapping function.

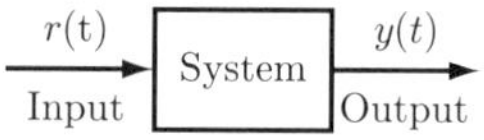

r(t)
Input
System
y(t)
Output

Figure 8.1 System

Let $r_1(t)$ and $r_2(t)$ be two signals, and a and b be nonzero real numbers. If $\mathcal{T}(\cdot)$ satisfies

$$\text{Additivity: } \mathcal{T}[r_1(t) + r_2(t)] = \mathcal{T}[r_1(t)] + \mathcal{T}[r_2(t)], \tag{8.2}$$

$$\text{Homogeneity: } \mathcal{T}[ar(t)] = a\mathcal{T}[r(t)], \tag{8.3}$$

or simply

Principle of superposition:

$$\mathcal{T}[ar_1(t) + br_2(t)] = a\mathcal{T}[r_1(t)] + b\mathcal{T}[r_2(t)], \tag{8.4}$$

the system is *linear* (線形). Note that, if the input is zero, then the corresponding output of a linear system is also zero because

$$\mathcal{T}(0) = \mathcal{T}(0 \cdot 0) = 0 \cdot \mathcal{T}(0) = 0. \tag{8.5}$$

The system output to a unit impulse input, $\delta(t)$, is called an *impulse response* (インパルス応答), which is given by

$$g(t) = \mathcal{T}[\delta(t)]. \tag{8.6}$$

The impulse response of a linear system is essential because it can be used to calculate the output of a linear system for any input signal. In fact, the output to an input $r(t)$ is

$$y(t) = g(t) * r(t) = \int_{-\infty}^{\infty} g(t-\tau)r(\tau)\mathrm{d}\tau, \tag{8.7}$$

that is, the output is the convolution integral of the impulse response and the input. Let $R(\omega) = \mathcal{F}[r(t)]$, $G(\omega) = \mathcal{F}[g(t)]$, and $Y(\omega) = \mathcal{F}[y(t)]$. The convolution theorem [(4.23) in Chapter 4] provides us with

$$Y(\omega) = G(\omega)R(\omega). \tag{8.8}$$

While the input-output relationship in the time domain, (8.7), is complex, the relationship in the frequency domain, (8.8), is simple. This simplicity shows the attractiveness of the Fourier transform. The output is simply obtained from $y(t) = \mathcal{F}^{-1}[Y(s)]$. An inverse Fourier transform can easily be calculated by looking up a Fourier transform table.

【Example 8.1】 Consider an RC circuit in **Figure 8.2** ($R = 2\,\Omega$ and $C = 1\,\mathrm{F}$), which is a first-order linear system

$$Ri(t) + v_C(t) = v_i(t),\ i(t) = C\frac{\mathrm{d}v_C(t)}{\mathrm{d}t}. \tag{8.9}$$

Find the impulse response of the system and calculate the output for the input $v_i(t) = f(t)$, where $f(t)$ is given in Example 4.1 for $\beta = 1$, that is,

$$f(t) = \begin{cases} 0, & \text{if } t < 0, \\ \mathrm{e}^{-t}, & \text{if } t \geqq 0. \end{cases} \tag{8.10}$$

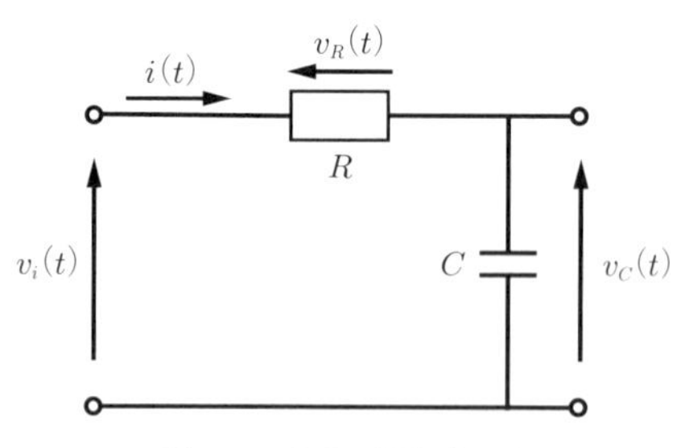

Figure 8.2 RC circuit

Substituting the values of R and C into (8.9) and performing the Fourier transform for the equation yield

$$\int_{-\infty}^{\infty}\left[2\frac{\mathrm{d}v_C(t)}{\mathrm{d}t} + v_C(t)\right]\mathrm{e}^{-j\omega t}\mathrm{d}t = \int_{-\infty}^{\infty} v_i(t)\mathrm{e}^{-j\omega t}\mathrm{d}t. \tag{8.11}$$

Let the Fourier transforms of $v_i(t)$ and $v_C(t)$ be $V_i(\omega)$ and $V_C(\omega)$, respectively; that is,

$$V_i(\omega) = \int_{-\infty}^{\infty} v_i(t)\mathrm{e}^{-j\omega t}\mathrm{d}t,\ V_C(\omega) = \int_{-\infty}^{\infty} v_C(t)\mathrm{e}^{-j\omega t}\mathrm{d}t. \tag{8.12}$$

The differential property provides us with $\int_{-\infty}^{\infty}\frac{\mathrm{d}v_C(t)}{\mathrm{d}t}\mathrm{e}^{-j\omega t}\mathrm{d}t = j\omega V_C(\omega)$. Thus, (8.11) becomes

$$j2\omega V_C(\omega) + V_C(\omega) = V_i(\omega). \tag{8.13}$$

When the input is chosen to be $v_i(t) = \delta(t)$, the impulse response of the system is given by

$$G(\omega) = \frac{V_C(\omega)}{V_i(\omega)} = \frac{1}{1+j2\omega}. \tag{8.14}$$

For the input signal (8.10), since its Fourier transform is $V_i(\omega) = \frac{1}{1+j\omega}$, (8.8) provides us with the response of the system

$$\begin{aligned} V_C(\omega) &= G(\omega)V_i(\omega) \\ &= \frac{1}{1+j2\omega}\frac{1}{1+j\omega} = \frac{1}{1/2+j\omega} - \frac{1}{1+j\omega}. \end{aligned} \tag{8.15}$$

Thus,

$$\begin{aligned} v_C(t) &= \mathcal{F}^{-1}[V_C(\omega)] = \mathcal{F}^{-1}\left(\frac{1}{1/2+j\omega} - \frac{1}{1+j\omega}\right) \\ &= \mathcal{F}^{-1}\left(\frac{1}{1/2+j\omega}\right) - \mathcal{F}^{-1}\left(\frac{1}{1+j\omega}\right) \\ &= \mathrm{e}^{-t/2} - \mathrm{e}^{-t}. \end{aligned} \tag{8.16}$$

8.2 Partial Differential Equation

An equation involving two or more independent variables $(x, y, \ldots)$, unknown functions $[u = u(x, y, \ldots), v = v(x, y, \ldots)]$, and their partial derivatives $(\partial u/\partial x, \partial^2 u/\partial t^2)$ is called a *partial differential equation* (偏微分方程式). Physical variables are functions of spatial coordinates and time, their relationships are usually described in partial differential equations in various mathematical and engineering problems.

The order of a partial differential equation is the highest order of the partial derivatives in the equation. Some important partial differential equations are

Heat equations

- One-dimensional: $\dfrac{\partial u}{\partial t} = c^2 \dfrac{\partial^2 u}{\partial^2 x}$
- Two-dimensional: $\dfrac{\partial u}{\partial t} = c^2 \left(\dfrac{\partial^2 u}{\partial^2 x} + \dfrac{\partial^2 u}{\partial^2 y} \right)$
- Three-dimensional: $\dfrac{\partial u}{\partial t} = c^2 \left(\dfrac{\partial^2 u}{\partial^2 x} + \dfrac{\partial^2 u}{\partial^2 y} + \dfrac{\partial^2 u}{\partial^2 z} \right).$

Wave equations

- One-dimensional: $\dfrac{\partial^2 u}{\partial t^2} = c^2 \dfrac{\partial^2 u}{\partial^2 x}$
- Two-dimensional: $\dfrac{\partial^2 u}{\partial t^2} = c^2 \left(\dfrac{\partial^2 u}{\partial^2 x} + \dfrac{\partial^2 u}{\partial^2 y} \right)$
- Three-dimensional: $\dfrac{\partial^2 u}{\partial t^2} = c^2 \left(\dfrac{\partial^2 u}{\partial^2 x} + \dfrac{\partial^2 u}{\partial^2 y} + \dfrac{\partial^2 u}{\partial^2 z} \right).$

Laplace equations

- Two-dimensional: $\dfrac{\partial^2 u}{\partial^2 x} + \dfrac{\partial^2 u}{\partial^2 y} = 0$
- Three-dimensional: $\dfrac{\partial^2 u}{\partial^2 x} + \dfrac{\partial^2 u}{\partial^2 y} + \dfrac{\partial^2 u}{\partial^2 z} = 0.$

Two-dimensional Poisson equation: $\dfrac{\partial^2 u}{\partial^2 x} + \dfrac{\partial^2 u}{\partial^2 y} = f(x, y).$

A partial differential equation is said to be linear if the equation only contains a linear combination of the dependent variable u and all its partial derivatives. Otherwise, it is said to be a nonlinear partial differential equation.

【Example 8.2】 Derive an equation describing the small transverse vibrations of an elastic string such as a guitar string. Stretch the string along the x-axis and fix the ends at $x = 0$ and $x = L$ (**Figure 8.3**).

Assume that the string is homogeneous; that the gravity is negligible; and that the string performs only small transverse motions in the

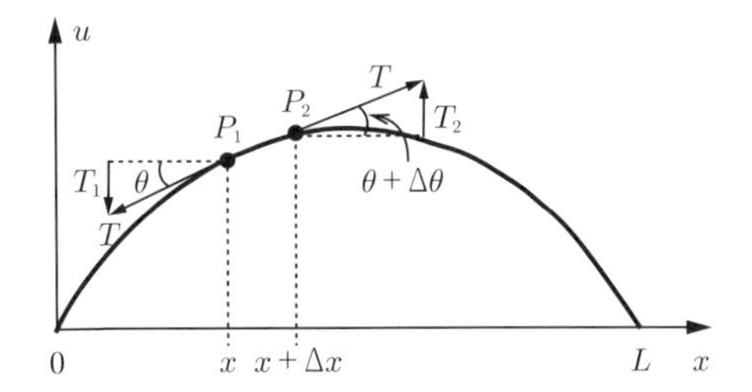

Figure 8.3 String vibration

vertical plane, that is, each point on the string has a small amount of movement strictly in the vertical direction.

Let the displacement of the string be $u(x,t)$. We establish a model for string vibrations. Since the tension at each point on the string is constant, we let it be T. We consider an infinitesimal part $[x, x+\Delta x]$ of the string and let ρ be the linear density. Each point on the string moves only vertically but not horizontally. The horizontal components of the tension are

$$T\cos(\theta+\Delta\theta) - T\cos\theta = 0, \tag{8.17}$$

and Newton's second law for this part in the vertical direction is

$$T\sin(\theta+\Delta\theta) - T\sin\theta = \rho\Delta x\frac{\partial^2 u}{\partial t^2}, \tag{8.18}$$

where $\rho\Delta x$ is the mass of the part, $\partial^2 u/\partial t^2$ is the vertical acceleration. Since $\theta \approx 0$, (8.17) yields

$$T\cos(\theta+\Delta\theta) = T\cos\theta \approx T. \tag{8.19}$$

Dividing (8.18) by (8.19) gives

$$\frac{T\sin(\theta+\Delta\theta)}{T\cos(\theta+\Delta\theta)} - \frac{T\sin\theta}{T\cos\theta} = \frac{\rho\Delta x}{T}\frac{\partial^2 u}{\partial t^2}, \tag{8.20}$$

that is,

$$\tan(\theta+\Delta\theta) - \tan\theta = \frac{\rho\Delta x}{T}\frac{\partial^2 u}{\partial t^2}. \tag{8.21}$$

Note that

$$\tan\theta = \left.\frac{\partial u}{\partial x}\right|_{x}, \quad \tan(\theta + \Delta\theta) = \left.\frac{\partial u}{\partial x}\right|_{x+\Delta x}. \tag{8.22}$$

Dividing (8.21) by Δx provides us with

$$\frac{1}{\Delta x}\left(\left.\frac{\partial u}{\partial x}\right|_{x+\Delta x} - \left.\frac{\partial u}{\partial x}\right|_{x}\right) = \frac{\rho}{T}\frac{\partial^2 u}{\partial t^2}. \tag{8.23}$$

Rewriting (8.23) yield the one-dimensional wave equation

$$\frac{\partial^2 u}{\partial t^2} = c^2\frac{\partial^2 u}{\partial x^2}, \quad c^2 = \frac{T}{\rho}. \tag{8.24}$$

A partial differential equation usually has many solutions. A unique solution can be found by using additional conditions, such as *boundary conditions* (境界条件) and *initial conditions* (初期条件).

If a partial differential equation is linear and homogeneous, then the *principle of superposition* (重ね合わせの原理) holds. Hence, if $u_1(t)$ and $u_2(t)$ are solutions of a homogeneous linear partial differential equation, then $u(t) = c_1u_1(t) + c_2u_2(t)$ with any constants c_1 and c_2 is also a solution of the equation.

The combination of the method of separating variables and a Fourier series is used to find a unique solution of a linear partial differential equation by the following steps:

Step 1) Use the method of separating variables to derive ordinary differential equations from a partial differential equation.

Step 2) Find the solution of each ordinary differential equation satisfying boundary conditions.

Step 3) Use a Fourier series to derive a unique solution that satisfies initial conditions and gives a solution for the entire problem.

【**Example 8.3**】 Find a unique solution for the one-dimensional wave equation (8.24), $u(x,t)$, for the boundary conditions

$$u(0,t) = 0,\ u(L,t) = 0,\quad \forall t \geqq 0, \tag{8.25}$$

and the initial conditions

$$u(x,0) = f(x),\quad \left.\frac{\partial u}{\partial t}\right|_{t=0} = g(x), \tag{8.26}$$

where $f(x)$ is the initial displacement and $g(x)$ is the initial velocity.

We use the above-mentioned three steps to solve this problem.

Step 1) Derive ordinary differential equations for the partial differential equation (8.24):

The method of separating variables is a powerful tool for solving engineering mathematical problems. It assumes that

$$u(x,t) = X(x)T(t) \tag{8.27}$$

is a solution to (8.24). This is called a product solution. Each function on the right side of (8.27) depends only on one of the variables, x and t.

Differentiating (8.27), we have

$$\frac{\partial^2 u}{\partial t^2} = X(x)\ddot{T}(t),\quad \frac{\partial^2 u}{\partial x^2} = X''(x)T(t), \tag{8.28}$$

where dots are derivatives with respect to t; and primes, to x.

Substituting (8.28) into (8.24) leads to

$$X(x)\ddot{T}(t) = c^2 X''(x)T(t), \tag{8.29}$$

or

$$\frac{\ddot{T}(t)}{c^2 T(t)} = \frac{X''(x)}{X(x)}. \tag{8.30}$$

Note that the terms on the left and right sides of (8.30) depend only on t and x, respectively. Thus, both sides must be constant. If we let k be a constant, then

$$\frac{\ddot{T}(t)}{c^2T(t)} = \frac{X''(x)}{X(x)} = k. \tag{8.31}$$

As a result, we obtain two ordinary differential equations:

$$\begin{cases} X''(x) - kX(x) = 0, \\ \ddot{T}(t) - c^2T(t) = 0. \end{cases} \tag{8.32}$$

Note that k is not yet determined at this step.

Step 2) Find the solutions satisfying boundary conditions:

The boundary conditions (8.25) becomes

$$u(0,t) = X(0)T(t) = 0,\ u(L,t) = X(L)T(t) = 0,\ \forall t \geqq 0. \tag{8.33}$$

Since $T(t) = 0$ is trivial, we assume that $T(t) \neq 0$. (8.33) yields

$$X(0) = 0,\ X(L) = 0,\ \ \forall t \geqq 0. \tag{8.34}$$

$k = 0$ in (8.32) results in $X(x) = ax + b$. And (8.34) yields a trivial solution $a = 0$ and $b = 0$. Thus, we assume that $k \neq 0$.

For $k > 0$, if we let $k = \mu^2$, then a general solution of $X(x)$ in (8.32) is given by

$$X(x) = \alpha \mathrm{e}^{\mu x} + \beta \mathrm{e}^{-\mu x}. \tag{8.35}$$

Applying the boundary conditions (8.34) provides us with $X(x) = 0$. It is also trivial. As a result, the possibility for a nontrivial solution is $k < 0$.

For $k < 0$, we let $k = -\omega^2$. Then, the first equation in (8.32) is

$$X''(x) + \omega^2 X(x) = 0 \tag{8.36}$$

and a general solution is

$$X(x) = \alpha \cos \omega x + \beta \sin \omega x. \tag{8.37}$$

The boundary conditions (8.34) gives

$$X(0) = \alpha = 0, \ X(L) = \beta \sin \omega L = 0. \tag{8.38}$$

Since $\beta = 0$ is trivial, we assume that $\beta \neq 0$. Thus, $\sin \omega L = 0$, which provides

$$\omega = \frac{n\pi}{L}, \ n = 1, 2, \ldots. \tag{8.39}$$

Note that the solution, $u(x,t)$, is a multiplication of the two items, $X(x)$ and $T(t)$, we can combine the gains to $T(t)$ and simply set $\beta = 1$. So, the general solution becomes

$$X(x) = \sin \frac{n\pi}{L} x, \ n = 1, 2, \ldots. \tag{8.40}$$

The second equation in (8.32) is

$$\ddot{T}(t) + \lambda_n^2 T(t) = 0, \ \lambda_n = \frac{cn\pi}{L}, \ n = 1, 2, \ldots. \tag{8.41}$$

In the same manner, we obtain a general solution of (8.41) as

$$T(t) = a_n \cos \lambda_n t + b_n \sin \lambda_n t. \tag{8.42}$$

As a result, the solution of (8.32) satisfying (8.25) is

$$u(x,t) = X(x)T(t) = \left(a_n \cos \frac{cn\pi}{L} t + b_n \sin \frac{cn\pi}{L} t \right) \sin \frac{n\pi}{L} x, \quad n = 1, 2, \ldots. \tag{8.43}$$

Step 3) Find a unique solution for the entire problem:

It is clear that $u(x,t)$ for a fixed n in (8.43) does not satisfy the initial conditions (8.26). It follows from the principle of superposition that the sum of $u(x,t)$ for all n is a solution of (8.43):

$$u(x,t) = \sum_{n=1}^{\infty} \left(a_n \cos \frac{cn\pi}{L} t + b_n \sin \frac{cn\pi}{L} t \right) \sin \frac{n\pi}{L} x. \tag{8.44}$$

The initial conditions (8.26) provide us with

$$u(x,0) = \sum_{n=1}^{\infty} a_n \sin \frac{n\pi}{L} x = f(x), \tag{8.45}$$

$$\left. \frac{\partial u}{\partial t} \right|_{t=0} = \sum_{n=1}^{\infty} b_n \frac{cn\pi}{L} \sin \frac{n\pi}{L} x = g(x). \tag{8.46}$$

Thus, if we choose a_n in (8.45) such that $u(x,0)$ is the Fourier series of $f(x)$, then we can determine a_n as

$$a_n = \frac{2}{L} \int_0^L f(x) \sin \frac{n\pi x}{L} \mathrm{d}x, \ n = 1, 2, \ldots. \tag{8.47}$$

Likewise, if we choose b_n in (8.46) such that $\partial u/\partial t|_{t=0}$ is the Fourier series of $g(x)$, then we can determine b_n as

$$b_n = \frac{2}{cn\pi} \int_0^L g(x) \sin \frac{n\pi x}{L} \mathrm{d}x, \ n = 1, 2, \ldots. \tag{8.48}$$

Problems

Basic Level

【1】 Let $R = 10\,\Omega$, $L = 1\,\mathrm{H}$, $M = 1\,\mathrm{kg}$, and $K = 0.1\,\mathrm{N/m}$. Calculate the system response of the systems in **Figure 8.4** for the following inputs

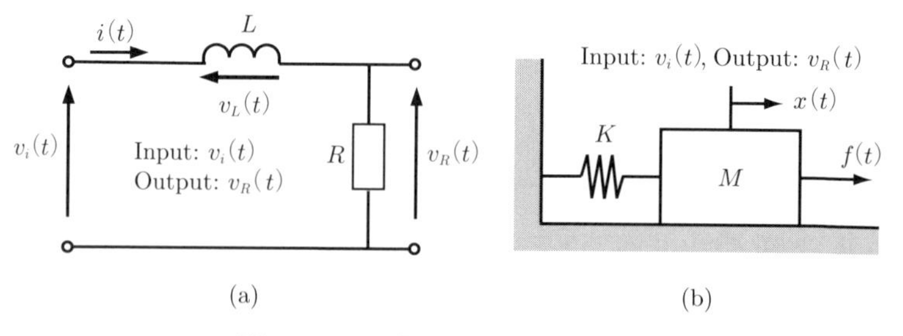

Figure 8.4 Basic-Level Problem 1

A. the unit impulse signal and

B. that has the form of (8.10).

【2】 Let the impulse response of a system be $G(\omega) = Ke^{-j\omega t_0}$. Show that the output of the system is given by $y(t) = Kr(t - t_0)$, where $r(t)$ and $y(t)$ are the input and output of the system, respectively.

【3】 Find a unique solution to a one-dimensional heat equation $\dfrac{\partial u}{\partial t} = \dfrac{\partial^2 u}{\partial x^2}$ $(0 \leqq x \leqq 1,\ t \geqq 0)$ for the boundary conditions $u(0,t) = 0$ and $u(1,t) = 0$, and the initial conditions $u(x,0) = 5,\ (0 \leqq x \leqq 1/2)$ and $u(x,0) = 0\ (1/2 < x \leqq 1)$.

【4】 Find a unique solution to a two-dimensional Laplace equation

$$\frac{\partial^2 u}{\partial x^2} + \frac{\partial^2 u}{\partial y^2} = 0,\ \ 0 \leqq x, y \leqq 1. \tag{8.49}$$

for the boundary conditions $u(x,0) = 0$, $u(x,1) = 0$, $u(0,y) = 1$, and $u(1,y) = 0$.

Advanced Level

【1】 Let $R = 1\,\Omega$, $L = 1\,\mathrm{H}$, $C = 1\,\mathrm{F}$, $M = 1\,\mathrm{kg}$, $K = 1\,\mathrm{N/m}$, and $b = \sqrt{2}\,\mathrm{Ns/m}$. Calculate the system response of the system in **Figure 8.5** for the following inputs

A. the unit step signal and

B. a sine wave $\sin t$ for $t \geqq 0$.

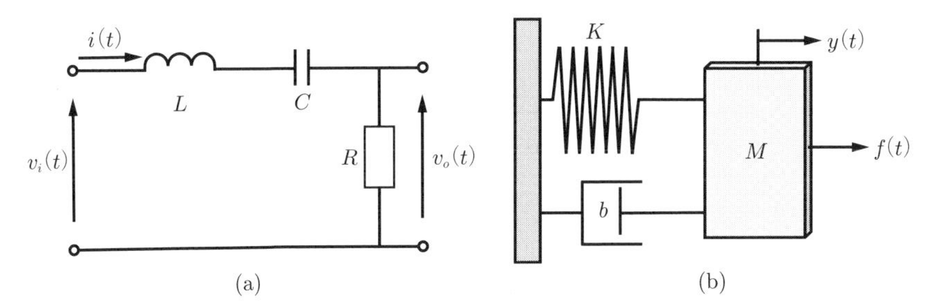

Figure 8.5 Advanced-Level Problem 1

【2】 Let $i(t)$ in Figure 8.4(a) be the output of the system and define the average power input to be $P = \dfrac{1}{T}\displaystyle\int_{-T/2}^{T/2} v_i(t)i(t)\mathrm{d}t$. Show the following facts:

A. $i(t)$ for an input $v_i(t) = V_0 + \sum_{n=1}^{\infty} V_n \cos n\omega t$ is given by $i(t) = I_0 + \sum_{n=1}^{\infty} I_n \cos(n\omega t + \phi_n)$.

B. $P = V_0 I_0 + \dfrac{1}{2} \sum_{n=1}^{\infty} V_n I_n \cos \phi_n$.

【3】 Find a unique solution for a two-dimensional wave equation

$$\frac{\partial^2 u(x,y,t)}{\partial t^2} = c^2 \left[\frac{\partial^2 u(x,y,t)}{\partial x^2} + \frac{\partial^2 u(x,y,t)}{\partial y^2} \right], \tag{8.50}$$

where a, b, and c are positive constants for $0 \leq x \leq a$, $0 \leq y \leq b$, and $t \geq 0$; the boundary conditions $u(0,y,t) = u(a,y,t) = u(x,0,t) = u(x,b,t) = 0$; and the initial conditions $u(x,y,0) = f(x,y)$ and $\left. \dfrac{\partial(x,y,t)}{\partial t} \right|_{t=0} = g(x,y)$. This wave equation describes the mathematical model of a rectangular membrane vibration (**Figure 8.6**), where $u(x,y,t)$ is a vertical deviation of the membrane orthogonal to the xy plane. The constant c is related to the membrane tension and mass per small unit area. $f(x,y)$ and $g(x,y)$ are the initial deviation and velocity of the membrane, respectively.

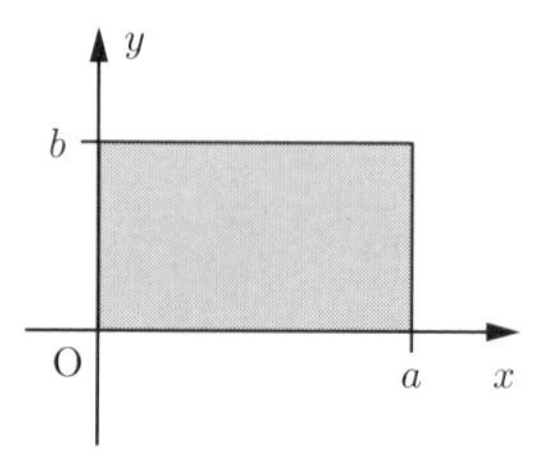

Figure 8.6 Rectangular membrane

【4】 Find a unique solution of a one-dimensional inhomogeneous Helmholtz equation $\dfrac{\mathrm{d}^2\phi(x)}{\mathrm{d}x^2} + k^2\phi(x) = \delta(x - x_0)$, which is defined on $-\infty < x < \infty$.

9 Multi-Dimensional Fourier Transform

We have discussed how to examine a time-domain signal and understand its frequency characteristics. Note that the number of the dimensions of many signals is usually larger than one, for example, an image has two or three dimensions (2-D or 3-D). The physical world has four dimensions [space (3-D) and time (1-D)]. Multi-dimensional phenomena have rich information. Performing a Fourier transform for such a signal provides us with many possibilities, just as we have seen for a 1-D signal. This chapter explains the extension of Fourier transforms to multi-dimensional signals, which may provide a deeper understanding of what is happening in nature than mere geometry.

9.1 Definition of Multi-Dimensional Fourier Transform

The Fourier transform of a 1-D signal, $f(t)$, is defined to be

$$F(\omega) = \mathcal{F}[f(t)] = \int_{-\infty}^{\infty} f(t)\mathrm{e}^{-j\omega t}\mathrm{d}t \tag{9.1}$$

and the inverse Fourier transform of the signal is given by

$$f(t) = \mathcal{F}^{-1}[F(\omega)] = \frac{1}{2\pi}\int_{-\infty}^{\infty} F(\omega)\mathrm{e}^{j\omega t}\mathrm{d}\omega. \tag{9.2}$$

Extending the definitions to a 2-D signal, $f(x_1, x_2)$, yields

$$\begin{aligned} F(\omega_1, \omega_2) &= \mathcal{F}[f(x_1, x_2)] \\ &= \int_{-\infty}^{\infty}\int_{-\infty}^{\infty} f(x_1, x_2)\mathrm{e}^{-j(\omega_1 x_1 + \omega_2 x_2)}\mathrm{d}x_1\mathrm{d}x_2 \\ &= \int_{-\infty}^{\infty}\int_{-\infty}^{\infty} f(x_1, x_2)\mathrm{e}^{-j\omega_1 x_1}\mathrm{e}^{-j\omega_2 x_2}\mathrm{d}x_1\mathrm{d}x_2 \end{aligned} \tag{9.3}$$

and

$$\begin{aligned} f(x_1, x_2) &= \mathcal{F}^{-1}[F(\omega_1, \omega_2)] \\ &= \frac{1}{(2\pi)^2} \int_{-\infty}^{\infty} \int_{-\infty}^{\infty} F(\omega_1, \omega_2) \mathrm{e}^{j(\omega_1 x_1 + \omega_2 x_2)} \mathrm{d}\omega_1 \mathrm{d}\omega_2 \\ &= \frac{1}{(2\pi)^2} \int_{-\infty}^{\infty} \int_{-\infty}^{\infty} F(\omega_1, \omega_2) \mathrm{e}^{j\omega_1 x_1} \mathrm{e}^{j\omega_2 x_2} \mathrm{d}\omega_1 \mathrm{d}\omega_2. \end{aligned} \tag{9.4}$$

To consider a general case, we let

$$x = [x_1, x_2, \ldots, x_n]^{\mathrm{T}} \tag{9.5}$$

be an n-dimensional (n-D) variable and $f(x)$ be a function of x. The Fourier and inverse Fourier transforms of $f(x)$ are given by†

$$\begin{aligned} &F(\omega_1, \omega_2, \ldots, \omega_n) = \mathcal{F}[f(x_1, x_2, \ldots, x_n)] \\ &= \int_{\mathbb{R}^n} f(x_1, x_2, \ldots, x_n) \mathrm{e}^{-j(\omega_1 x_1 + \omega_2 x_2 + \cdots + \omega_n x_n)} \mathrm{d}x_1 \mathrm{d}x_2 \cdots \mathrm{d}x_n \end{aligned} \tag{9.6}$$

and

$$\begin{aligned} &f(x_1, x_2, \ldots, x_n) = \mathcal{F}^{-1}[F(\omega_1, \omega_2, \ldots, \omega_n)] \\ &= \frac{1}{(2\pi)^n} \int_{\mathbb{R}^n} F(\omega_1, \omega_2, \cdots, \omega_n) \mathrm{e}^{j(\omega_1 x_1 + \omega_2 x_2 + \cdots + \omega_n x_n)} \\ &\times \mathrm{d}\omega_1 \mathrm{d}\omega_2 \cdots \mathrm{d}\omega_n, \end{aligned} \tag{9.7}$$

or simply

$$F(\omega) = \mathcal{F}[f(x)] = \int_{\mathbb{R}^n} f(x) \mathrm{e}^{-j\omega \cdot x} \mathrm{d}x \tag{9.8}$$

and

$$f(x) = \mathcal{F}^{-1}[F(\omega)] = \frac{1}{(2\pi)^n} \int_{\mathbb{R}^n} F(\omega) \mathrm{e}^{j\omega \cdot x} \mathrm{d}\omega, \tag{9.9}$$

where

† $\mathbb{R}^n$ is a real coordinate space of dimension n.

$$\begin{cases} \omega \cdot x = \omega_1 x_1 + \omega_2 x_2 + \cdots + \omega_n x_n, \\ \mathrm{d}x = \mathrm{d}x_1 \mathrm{d}x_2 \cdots \mathrm{d}x_n, \\ \mathrm{d}\omega = \mathrm{d}\omega_1 \mathrm{d}\omega_2 \cdots \mathrm{d}\omega_n. \end{cases} \tag{9.10}$$

Thus, if we use a vector notation, we can describe the n-D Fourier transform like the 1-D Fourier transform.

The properties of the n-D Fourier transform are basically the same as those of the 1-D Fourier transform discussed in previous chapters. Nevertheless, it is useful to collect some basic facts below. In addition to† (9.5), we assume that

$$\omega = [\omega_1, \omega_2, \ldots, \omega_n]^{\mathrm{T}}. \tag{9.11}$$

Linearity:

$$\mathcal{F}[\alpha f(x) + \beta g(x)] = \alpha \mathcal{F}[f(x)] + \beta \mathcal{F}[g(x)] = \alpha F(\omega) + \beta G(\omega), \tag{9.12}$$

where α and β are constants.

Translation:

$$\mathcal{F}[f(x \pm x_0)] = \mathrm{e}^{\pm j\omega \cdot x_0} \mathcal{F}[f(x)] = \mathrm{e}^{\pm j\omega \cdot x_0} F(\omega), \tag{9.13}$$

where x_0 is a shifted displacement.

Convolution: A 2-D version is

$$f(x_1, x_2) * g(y_1, y_2) = \int_{-\infty}^{\infty} \int_{-\infty}^{\infty} f(x_1 - y_1, x_2 - y_2) g(y_1, y_2) \mathrm{d}y_1 \mathrm{d}y_2 \tag{9.14}$$

and its Fourier transform is

$$\mathcal{F}[f(x_1, x_2) * g(y_1, y_2)] = \mathcal{F}[f(x_1, x_2)] \mathcal{F}[g(y_1, y_2)]$$

† In addition to ～：～に加えて，～のほかに。

$$= F(\omega_1, \omega_2)G(\omega_1, \omega_2). \tag{9.15}$$

A general version is

$$f(x) * g(y) = \int_{\mathbb{R}^n} f(x-y)g(y)\mathrm{d}y \tag{9.16}$$

and

$$\mathcal{F}[f(x) * g(y)] = \mathcal{F}[f(x)]\mathcal{F}[g(y)] = F(\omega)G(\omega). \tag{9.17}$$

Stretch: A 2-D version is

$$\mathcal{F}[f(a_1x_1, a_2x_2)] = \frac{1}{|a_1||a_2|}F\left(\frac{\omega_1}{a_1}, \frac{\omega_2}{a_2}\right), \tag{9.18}$$

where a_1 and a_2 are constants. A general version is

$$\mathcal{F}[f(Ax)] = \frac{1}{|\det A|}F(A^{-\mathrm{T}}\omega), \tag{9.19}$$

where A is a transformation matrix.

Same as that for the 1-D case, we define the unit impulse signal to be

$$\begin{cases} \delta(x_1, x_2, \ldots, x_n) = \begin{cases} \infty, & \text{if } x_1 = x_2 = \cdots = x_n = 0, \\ 0, & \text{otherwise}, \end{cases} \\ \displaystyle\int_{-\infty}^{\infty}\int_{-\infty}^{\infty}\cdots\int_{-\infty}^{\infty} \delta(x_1, x_2, \ldots, x_n)\mathrm{d}x_1\mathrm{d}x_2\cdots\mathrm{d}x_n = 1, \end{cases} \tag{9.20}$$

and the unit step signal as

$$1(x_1, x_2, \ldots, x_n) = \begin{cases} 1, & \text{if } x_1 \geqq 0, x_2 \geqq 0, \ldots, \text{ and } x_n \geqq 0, \\ 0, & \text{otherwise}. \end{cases} \tag{9.21}$$

9.2 Application to Image Compression

An image is composed of pixels. Assuming that pixels are arranged in a square-lattice form, just like graph paper. If an image is divided by N rows and M columns, then the image is described by $N \times M$ pixels.

A *gray level* (濃度値, 階調値, 輝度値) is used to describe the intensity (or brightness) of each pixel in a monochrome image. If monochrome is quantized into n gray levels (for example, **Figure 9.1**), the gray level of each lattice is given by, for example, the mean value of gray levels at four-corner positions or those in the whole lattice area. An $N \times M$ gray-level matrix provides us with the coding result of the image. The gray levels are usually set to be 256 ($= 2^8$) for a normal image, but are set to be 1024 ($= 2^{10}$) or 4096 ($= 2^{12}$) for medical images and remote sensing.

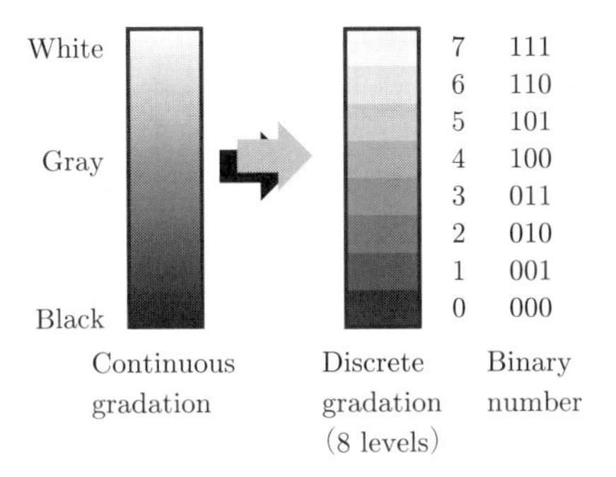

Figure 9.1 8-bit quantization of gradation

Several methods have been presented to quantize a color. For example, the RGB color method decomposes the color of a pixel into three primary colors: red, green, and blue. Since each color can be treated as an independent variable, we quantize gray levels for each of the colors of a color image as for a monochrome image and obtain the coding data for a color image (**Figure 9.2**).

Now, we explain the *discrete cosine transform* (DCT, 離散コサイン変換). While the FFT of a real number series is complex, its DCT is real. Moreover, the DCT has strong energy compaction, that is, it features that

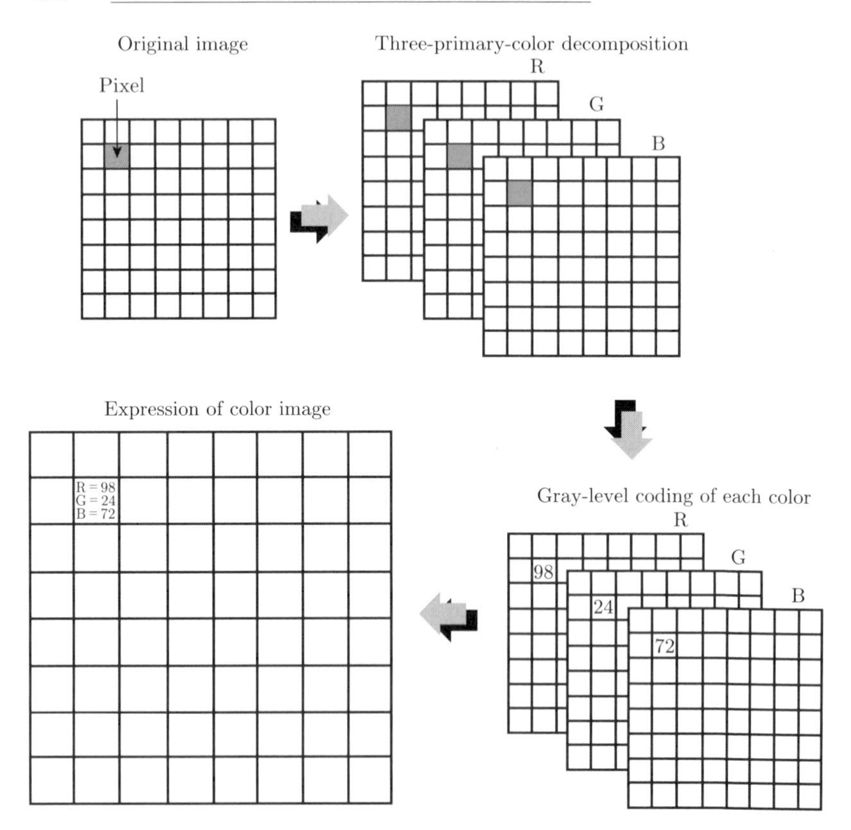

Figure 9.2 Coding of color image

the information is stored in low frequencies. For simplicity, we use a 1-D signal to explain the basic idea.

As shown in (6.8), the DFT of a signal $f(t)$ is given by

$$F_n = \sum_{m=0}^{N-1} f_m e^{-j\frac{2\pi nm}{N}}, \ n = 0, 1, \ldots, N-1, \tag{9.22}$$

where $f_m = f(m\Delta T)$ $(m = 0, 1, \ldots, N)$, ΔT is a sampling period, and N is the number of points for DFT. If we choose the points to be [**Figure 9.3**(a)]

$$f_m = f\left(m\Delta T + \frac{\Delta T}{2}\right), \ m = 0, 1, \ldots, N-1, \tag{9.23}$$

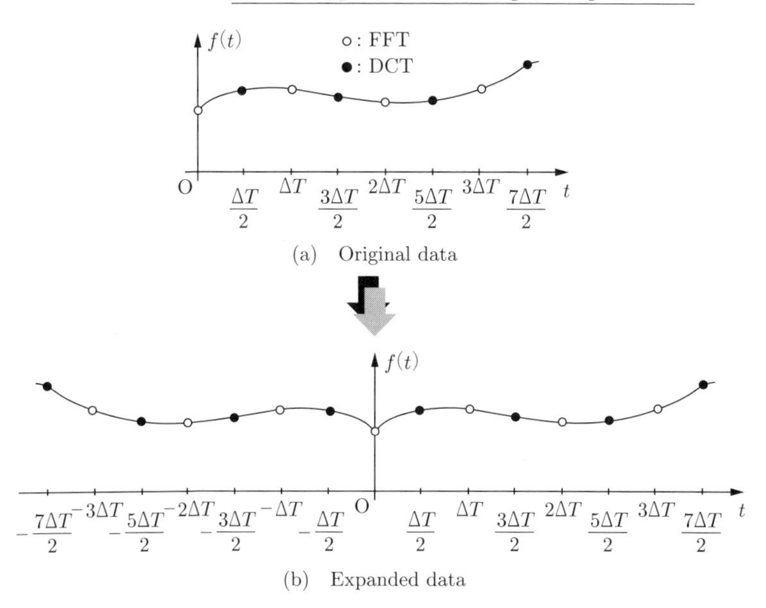

Figure 9.3 Discrete cosine transform

expending the signal by wrapping the signal about the vertical axis [Figure 9.3(b)] and repeating the data every $2N$ points (period: $2N\Delta T$), we obtain an even periodic function $f_T(t)$ that is expressed as [see (3.31)]

$$f_T(t) = \frac{a_0}{2} + \sum_{n=1}^{\infty} a_n \cos n\omega t, \ \omega = \frac{\pi}{N\Delta T}. \tag{9.24}$$

Since we only have N points of the original data, we choose the first N terms and approximate (9.24) to be

$$f_m = f\left(m\Delta T + \frac{\Delta T}{2}\right) = \frac{a_0}{2} + \sum_{n=1}^{N-1} a_n \cos \frac{n(2m+1)\pi}{2N},$$
$$m = 0, 1, \ldots, N-1. \tag{9.25}$$

Based on (3.19),

$$a_n = \frac{2}{T}\int_{-T/2}^{T/2} f_T(t) \cos n\omega \mathrm{d}t = \frac{4}{T}\int_0^{T/2} f_T(t) \cos n\omega \mathrm{d}t$$

$$\approx \frac{2}{N}\sum_{m=0}^{N-1} f_m \cos\frac{(2m+1)n\pi}{2N},\ n = 0, 1, \ldots, N-1. \tag{9.26}$$

Thus, the DCT of f_m in (9.25) is

$$F_{c0} = \frac{1}{2}\sum_{m=0}^{N-1} f_m,\ F_{cn} = \sum_{m=0}^{N-1} f_m \cos\frac{(2m+1)n\pi}{2N},$$
$$n = 0, 1, \ldots, N-1. \tag{9.27}$$

Letting

$$\alpha_0 = \frac{a_0}{2} = \frac{2}{N}F_{c0},\ \alpha_n = a_n = \frac{2}{N}F_{cn},\ n = 1, 2, \ldots, N-1 \tag{9.28}$$

allows us to write (9.24) as

$$f_m = \sum_{n=0}^{N-1} \alpha_n \cos\frac{(2m+1)n\pi}{2N},\ m = 1, 2, \ldots, N-1. \tag{9.29}$$

We define (9.27) to be the DCT of f_m and (9.29) to be its *inverse discrete cosine transform* (IDCT, 逆離散コサイン変換)[†].

The DCT of a 2-D function $f(x, y)$ is

$$F_c(k_1, k_2)$$
$$= a(k_1)b(k_2)\sum_{x=0}^{N_1-1}\sum_{y=0}^{N_2-1} f(x, y)\cos\frac{k_1(2x+1)\pi}{2N_1}\cos\frac{k_2(2y+1)\pi}{2N_2},$$
$$k_1 = 0, 1, \ldots, N_1-1,\ k_2 = 0, 1, \ldots, N_2-1, \tag{9.30}$$

and its IDCT is

$$f(x, y) = \frac{4}{N_1N_2}\sum_{x=0}^{N_1-1}\sum_{y=0}^{N_2-1} a(k_1)b(k_2)F_c(k_1, k_2)$$
$$\times\cos\frac{k_1(2x+1)\pi}{2N_1}\cos\frac{k_2(2y+1)\pi}{2N_2},$$
$$x = 0, 1, \ldots, N_1-1,\ y = 0, 1, \ldots, N_2-1, \tag{9.31}$$

where

[†] The coefficient $2/N$ is included in the IDCT in this definition. Including $\sqrt{2}/\sqrt{N}$ in both DCT and IDCT gives another definition but the true nature of the definitions is the same.

$$a(k_1) = \begin{cases} \dfrac{1}{\sqrt{2}}, & \text{if} \quad k_1 = 0, \\ 1 & \text{if} \quad k_1 = 1, 2, \ldots, N_1 - 1, \end{cases}$$

$$b(k_2) = \begin{cases} \dfrac{1}{\sqrt{2}}, & \text{if} \quad k_2 = 0, \\ 1 & \text{if} \quad k_2 = 1, 2, \ldots, N_2 - 1. \end{cases}$$

The DCT is used for information compression. A typical example is the jpeg (joint photographic experts group) method that is commonly used for the lossy compression of digital images. The procedure has two steps: First, divide an image into blocks with 8×8 pixels and calculate the gray levels

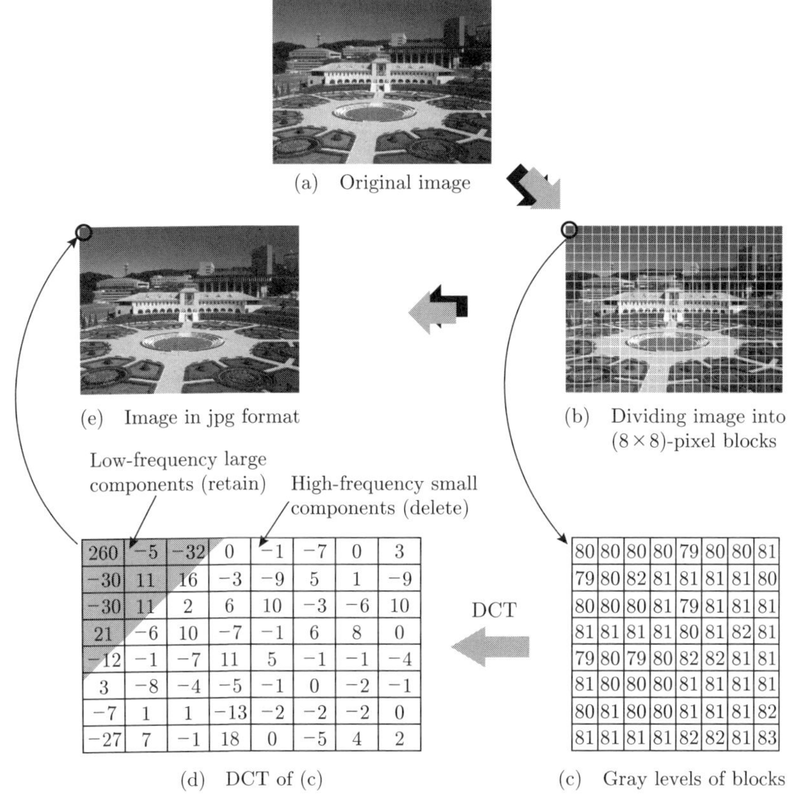

260	−5	−32	0	−1	−7	0	3
−30	11	16	−3	−9	5	1	−9
−30	11	2	6	10	−3	−6	10
21	−6	10	−7	−1	6	8	0
−12	−1	−7	11	5	−1	−1	−4
3	−8	−4	−5	−1	0	−2	−1
−7	1	1	−13	−2	−2	−2	0
−27	7	−1	18	0	−5	4	2

(d) DCT of (c)

80	80	80	80	79	80	80	81
79	80	82	81	81	81	81	80
80	80	80	81	79	81	81	81
81	81	81	81	80	81	82	81
79	80	79	80	82	82	81	81
81	80	80	80	81	81	81	81
80	81	80	80	81	81	81	82
81	81	81	81	82	82	81	83

(c) Gray levels of blocks

Figure 9.4 Example of DCT

of blocks [**Figure 9.4**(b) and (c)]. Then, perform the DCT for the image. Since the DCT collects the components at low frequencies, we reduce the size of the image by discarding high-frequency components [Figure 9.4(d) and (e)].

While the jpg format is effective for many images, it is difficult to deal with images having clear contours, and the png (portable network graphic) format is recommended for those images.

9.3 Application to Computerized Tomography

CT is a technology that uses *X-ray anthropometry* (X 線人体測定) to construct the internal image of a human body through image processing.

If we set a source and a detector of X-rays on each side of a human body, X-rays pass through and are attenuated by the body, and are then received by the detector on the opposite side (**Figure 9.5**).

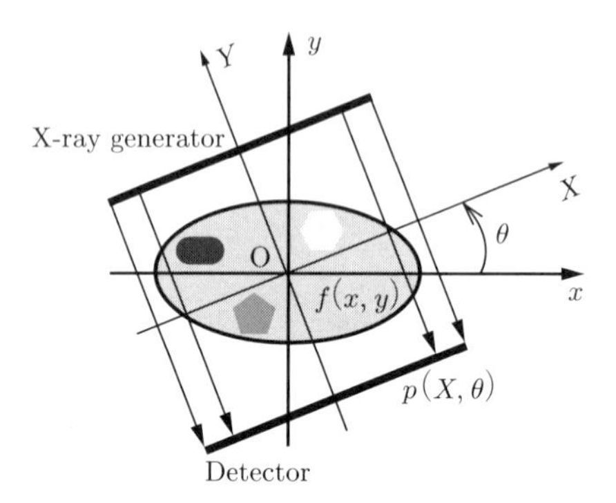

Figure 9.5 2-D parallel-beam projection.

The distribution of measurements changes due to the difference in X-ray absorption at each position inside a human body, thus obtaining an X-ray image. The pair of the X-ray source and detector rotate little by little[†1], continue recording images, and perform the Fourier transforms for the images.

Finally, we perform an inverse Fourier transform and obtain a gray CT image that shows the amount of X-rays absorbed at each position inside the body in brightness from black to white.

The principle of the reconstruction of a two-dimensional X-ray CT image is explained as follows.

Let xy be a world coordinate system[†2]. Rotating the xy system by θ degrees yields an XY coordinate system. The relationship between these two systems are

$$\begin{cases} X = x\cos\theta + y\sin\theta, \\ Y = -x\sin\theta + y\cos\theta. \end{cases} \tag{9.32}$$

Let the intensity of the X-ray beam at the incident side be I_i, the intensity after transmission through an object be I_o, and the attenuation distribution for the X-ray in a cross-section of the object be $f(x, y)$. Then,

$$I_o = I_i \mathrm{e}^{-\int_{-\infty}^{\infty} f(x,y)\mathrm{d}Y}. \tag{9.33}$$

The corresponding projection data is

$$\begin{aligned} p(X,\theta) &= \ln\frac{I_i}{I_o} = \int_{-\infty}^{\infty} f(x,y)\mathrm{d}Y \\ &= \int_{-\infty}^{\infty} f(X\cos\theta - Y\sin\theta, X\sin\theta + Y\cos\theta)\mathrm{d}Y, \end{aligned} \tag{9.34}$$

which is called the Radon transform.

We collect $p(X, \theta)$ over the entire circumference of the object ($0 \leq \theta \leq 2\pi$) and compute $f(x, y)$ based on the projection data.

---

†1 little by little：少しずつ。

†2 a world coordinate system：ワールド座標系。扱っている空間全体の座標系であり，中にある個別の物体にそれぞれローカル座標系（local coordinate system）を別途に設定することにより，全体空間の中における各物体の変化を扱いやすくする。

The Fourier transform of $p(X,\theta)$ is

$$\begin{aligned}
P(\rho,\theta) &= \int_{-\infty}^{\infty} p(X,\theta)\mathrm{e}^{-j\rho X}\mathrm{d}X \\
&= \int_{-\infty}^{\infty}\left(\int_{-\infty}^{\infty} f(x,y)\mathrm{d}Y\right)\mathrm{e}^{-j\rho X}\mathrm{d}X \\
&= \int_{-\infty}^{\infty}\int_{-\infty}^{\infty} f(x,y)\mathrm{e}^{-j\rho(x\cos\theta+y\sin\theta)}\mathrm{d}X\mathrm{d}Y \\
&= \int_{-\infty}^{\infty}\int_{-\infty}^{\infty} f(x,y)\mathrm{e}^{-j(\rho\cos\theta)x-j(\rho\sin\theta)y}\mathrm{d}x\mathrm{d}y \\
&= F(\rho\cos\theta,\rho\sin\theta),
\end{aligned} \tag{9.35}$$

where $\mathrm{d}X\mathrm{d}Y = \mathrm{d}x\mathrm{d}y$. If we assume that

$$u = \rho\cos\theta, \quad v = \rho\sin\theta, \tag{9.36}$$

it is clear from the last equation in (9.35) that the Fourier transform of $f(x,y)$, $F(u,v)$, is exactly the Fourier transform of $p(X,\theta)$.

The inverse Fourier transform of $F(u,v)$ is

$$\begin{aligned}
f(x,y) &= \frac{1}{4\pi^2}\int_{-\infty}^{\infty} F(u,v)\mathrm{e}^{j(ux+vy)}\mathrm{d}u\mathrm{d}v \\
&= \frac{1}{4\pi^2}\int_{0}^{2\pi}\int_{0}^{\infty} F(\rho\cos\theta,\rho\sin\theta)\mathrm{e}^{j\rho(x\cos\theta+y\sin\theta)}\rho\mathrm{d}\rho\mathrm{d}\theta,
\end{aligned} \tag{9.37}$$

where $\mathrm{d}u\mathrm{d}v = \rho\mathrm{d}\rho\mathrm{d}\theta$. Moreover,

$$\begin{aligned}
f(x,y) &= \frac{1}{4\pi^2}\int_{0}^{2\pi}\int_{0}^{\infty} P(\rho,\theta)\mathrm{e}^{j\rho(x\cos\theta+y\sin\theta)}\rho\mathrm{d}\rho\mathrm{d}\theta \\
&= \frac{1}{4\pi^2}\int_{0}^{2\pi}\left[\int_{0}^{\infty} P(\rho,\theta)\rho\mathrm{e}^{j\rho(x\cos\theta+y\sin\theta)}\mathrm{d}\rho\right]\mathrm{d}\theta.
\end{aligned} \tag{9.38}$$

Thus, taking the inverse Fourier transform of $P(\rho,\theta)\rho\times 1(t)$ and integrating it for every angle ($0 \leq \theta \leq 2\pi$) yield the attenuation distribution $f(x,y)$.

Problems

Basic Level

[1] Find the two-dimensional Fourier transform of the following functions:

A. $f(x, y) = \delta(x, y)$.

B. $f(x, y) = \cos 2\pi(ax + by)$, where a and b are constant.

C. $f(x, y) = \mathrm{e}^{-\pi(x^2+y^2)}$.

[2] Let $T = 3\Delta T$ and the original discrete signal is $\{x_0, x_1, x_2\}$ (**Figure 9.6**). It follows from (9.25) that the DST of the signal is

$$\begin{cases} x_0 = x\left(\dfrac{\Delta T}{2}\right) = a_0 + a_1 \cos \dfrac{\pi}{3\Delta T}\dfrac{\Delta T}{2} + a_2 \cos \dfrac{2\pi}{3\Delta T}\dfrac{\Delta T}{2}, \\ x_1 = x\left(\dfrac{3\Delta T}{2}\right) = a_0 + a_1 \cos \dfrac{\pi}{3\Delta T}\dfrac{3\Delta T}{2} + a_2 \cos \dfrac{2\pi}{3\Delta T}\dfrac{3\Delta T}{2}, \\ x_2 = x\left(\dfrac{5\Delta T}{2}\right) = a_0 + a_1 \cos \dfrac{\pi}{3\Delta T}\dfrac{5\Delta T}{2} + a_2 \cos \dfrac{2\pi}{3\Delta T}\dfrac{5\Delta T}{2}, \end{cases} \tag{9.39}$$

that is,

$$\begin{bmatrix} x_0 \\ x_1 \\ x_2 \end{bmatrix} = \begin{bmatrix} 1 & \cos \dfrac{\pi}{6} & \cos \dfrac{2\pi}{6} \\ 1 & \cos \dfrac{3\pi}{6} & \cos \dfrac{6\pi}{6} \\ 1 & \cos \dfrac{5\pi}{6} & \cos \dfrac{10\pi}{6} \end{bmatrix} \begin{bmatrix} a_0 \\ a_1 \\ a_2 \end{bmatrix}. \tag{9.40}$$

Solving (9.40) yields the DCT of $\{x_i\}$

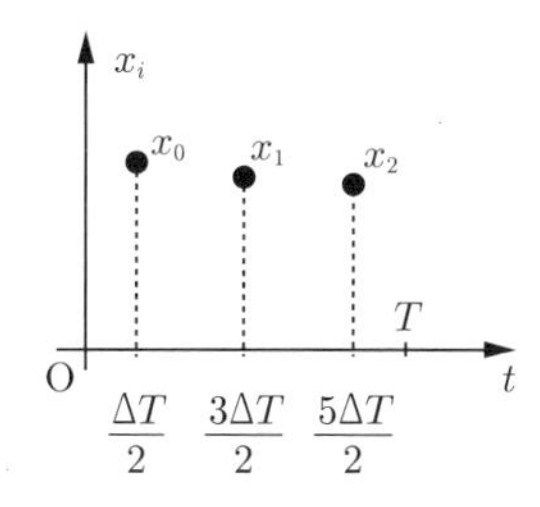

Figure 9.6 Discrete signal

$$\begin{bmatrix} a_0 \\ a_1 \\ a_2 \end{bmatrix} = \begin{bmatrix} \frac{1}{3} & \frac{1}{3} & \frac{1}{3} \\ \frac{2}{3}\cos\frac{\pi}{6} & \frac{2}{3}\cos\frac{3\pi}{6} & \frac{2}{3}\cos\frac{5\pi}{6} \\ \frac{2}{3}\cos\frac{2\pi}{6} & \frac{2}{3}\cos\frac{6\pi}{6} & \frac{2}{3}\cos\frac{10\pi}{6} \end{bmatrix} \begin{bmatrix} x_0 \\ x_1 \\ x_2 \end{bmatrix}. \tag{9.41}$$

Calculate the DCT of $\{x_i\} = \{1, 0.8, 0.6\}$.

【3】 Understand the explanation in Basic-Level Problem 2 and calculate the DCTs of the signals in Basic-Level Problem 3 in Chapter 6.

【4】 The basic problem in CT scans is image reconstruction. Let a 2-D image of 4 pixels be **Figure 9.7**(a). Emitting X-rays at 0° yields the result $\{4, 6\}$ [Figure 9.7(b)]; and at 90°, the result $\{3, 7\}$ [Figure 9.7(c)]. Then, we have to estimate $\{a, b, c, d\}$ in the 2-D image based on the projection results $\{4, 6\}$ and $\{3, 7\}$, that is, we have to solve

$$a + b = 4, \ \ c + d = 6, \ \ a + c = 3, \ \ b + d = 7, \tag{9.42}$$

or

$$Qx = y, \ Q = \begin{bmatrix} 1 & 1 & 0 & 0 \\ 0 & 0 & 1 & 1 \\ 1 & 0 & 1 & 0 \\ 0 & 1 & 0 & 1 \end{bmatrix}, \ x = \begin{bmatrix} a \\ b \\ c \\ d \end{bmatrix}, \ y = \begin{bmatrix} 4 \\ 6 \\ 3 \\ 7 \end{bmatrix} \tag{9.43}$$

to find $\{a, b, c, d\}$. Since Q in (9.43) is not invertible, we cannot directly solve (9.43) to obtain x. The Radon transform, (9.34), is then used in conjunction

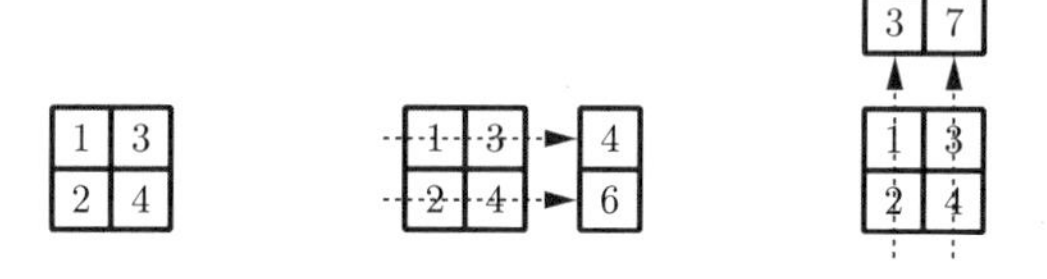

(a) 2-D image (b) Projection at 0° (c) Projection at 90°

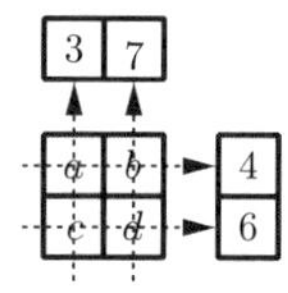

(d) Image reconstruction

Figure 9.7 Problem of image reconstruction

with the Fourier transform to find an optimal solution to this problem. For simplicity in this exercise, we use the Moore-Penrose pseudoinverse to find a solution. The MATLAB command is `pinv`. Calculate $\{a, b, c, d\}$ for the projection results $\{8, 16\}$ and $\{10, 14\}$.

【5】 Let the DCT of a picture with 8×8 pixels be (9.30), where $N_1 = N_2 = 8$. Explain how to reduce the size of the image by discarding high-frequency components.

Advanced Level

【1】 Various filters are used in image-data processing. Describe their purposes and characteristics.

【2】 There are two types of compression methods for image data: lossless compression and lossy compression. Describe the differences and typical compression methods for each.

10 Laplace Transform

The Laplace transform, which is named after Pierre-Simon Laplace, transforms a function in the time domain into a function in a complex frequency domain (called the s domain). It has a close relationship with the Fourier transform. Since the Laplace transform has a wider range of applicable functions and is easier to use than the Fourier transform, it has been widely used as a tool for solving differential equations, designing electric circuits, analyzing system stability, etc.

10.1 Definition of Laplace Transform

As shown in Chapter 4, a signal has to be absolutely integrable to ensure the existence of its Fourier transform. This condition is very strict. A step signal, a sine signal, and many other well-known signals do not satisfy this condition. We modify a signal $f(t)$ as $f(t)1(t)\mathrm{e}^{-\beta t}$ to make it easy to satisfy the absolute integrability condition and expand the range of applications of a Fourier transform. Since real-time signal processing may not care about signals before processing starts, multiplying a signal by the unit step signal, $1(t)$, changes the integral range from $(-\infty, \infty)$ to $[0, \infty)$. And multiplying a signal by the factor $\mathrm{e}^{-\beta t}$ decays it over time.

Performing the Fourier transform for the modified signal gives

$$F(\omega) = \int_{-\infty}^{\infty} [f(t)1(t)\mathrm{e}^{-\beta t}]\mathrm{e}^{-j\omega t}\mathrm{d}t = \int_{0}^{\infty} f(t)\mathrm{e}^{-(\beta+j\omega)t}\mathrm{d}t. \quad (10.1)$$

Letting $s = \beta + j\omega$ yields

$$F(s) = \int_0^\infty f(t)\mathrm{e}^{-st}\mathrm{d}t, \tag{10.2}$$

which is called the *Laplace transform* (ラプラス変換) of the signal $f(t)$. On the other hand, the *inverse Laplace transform* (ラプラス逆変換) is defined to be

$$f(t) = \frac{1}{2\pi j}\int_{c-j\infty}^{c+j\infty} F(s)\mathrm{e}^{st}\mathrm{d}s, \tag{10.3}$$

where c is a real number. The Laplace transform is commonly used in electrical and electronic engineering, mechanical engineering, control engineering, and many other fields.

Some properties of the Laplace transform are listed in **Table 10.1**, where

$$\begin{cases} F_i(s) = \mathcal{L}\{f_i(t)\}, \ i = 1, 2, \\ f(t) * g(t) = \displaystyle\int_0^t f(\tau)g(t-\tau)\mathrm{d}\tau, \end{cases} \tag{10.4}$$

and a and b are constants. Note that $f(t) = 0$ for $t < 0$.

Table 10.1 Properties of Laplace transform

Name	Formula
Linearity	$\mathcal{L}[af_1(t) + bf_2(t)] = aF_1(s) + bF_2(s)$
Differentiation	$\mathcal{L}\left[\frac{\mathrm{d}f(t)}{\mathrm{d}t}\right] = sF(s) - f(0)$ $\mathcal{L}\left[\frac{\mathrm{d}^n f(t)}{\mathrm{d}t^n}\right] = s^n F(s) - s^{n-1}f(0) - s^{n-2}\left.\frac{\mathrm{d}f(t)}{\mathrm{d}t}\right\|_{t=0} - \cdots - \left.\frac{\mathrm{d}^{(n-1)}f(t)}{\mathrm{d}t^{(n-1)}}\right\|_{t=0}$
Integration	$\mathcal{L}\left[\int_0^t f(\tau)d\tau\right] = \frac{1}{s}F(s)$ $\mathcal{L}\left[\underbrace{\int_0^t \mathrm{d}t\int_0^t \mathrm{d}t\cdots\int_0^t}_{n} f(t)\mathrm{d}t\right] = \frac{1}{s^n}F(s)$
Convolution	$\mathcal{L}[f_1(t) * f_2(t)] = F_1(s)F_2(s)$

10.2 Frequency Response

When we add a signal to a system, The system response can be divided into two parts: the *transient response* (過渡応答) and the *steady-state response* (定常応答). Since the superposition principle holds for a linear system, we can decompose a complex signal into the combination of sine waves with different angular frequencies and verify the system response for each sinusoidal input. This provides us with a tool called the *frequency response* (周波数応答) for system analysis.

The response of a linear system to a sinusoidal input, $v_i \sin \omega t$, in the steady state is a sine wave with the same angular frequency but different amplitude and phase, $v_o \sin(\omega t + \phi)$, which gives valuable information about the behavior of the system in the frequency domain. The Fourier-transform method can be used to calculate such a steady-state response.

The input-output relationship of a linear system is given by an ordinary differential equation

$$\begin{aligned}&\frac{\mathrm{d}^n y(t)}{\mathrm{d}t^n} + a_{n-1}\frac{\mathrm{d}^{n-1} y(t)}{\mathrm{d}t^{n-1}} + \cdots + a_1 \frac{\mathrm{d}y(t)}{\mathrm{d}t} + a_0 y(t) \\ &= b_m \frac{\mathrm{d}^m u(t)}{\mathrm{d}t^m} + b_{m-1}\frac{\mathrm{d}^{m-1} u(t)}{\mathrm{d}t^{m-1}} + \cdots + b_1 \frac{\mathrm{d}u(t)}{\mathrm{d}t} + b_0 u(t),\end{aligned} \tag{10.5}$$

where $u(t)$ and $y(t)$ are the input and output of the system, respectively; and a_i and b_j $(i = 0, 1, 2, \ldots, n; j = 0, 1, 2, \ldots, m)$ are constants that determine system characteristics.

The Fourier transform of the system (10.5) yields

$$\begin{aligned}&(j\omega)^n Y(j\omega) + a_{n-1}(j\omega)^{n-1} Y(j\omega) + \cdots + a_1(j\omega)Y(j\omega) + a_0 Y(j\omega) \\ &= b_m (j\omega)^m U(j\omega) + b_{m-1}(j\omega)^{m-1} U(j\omega) + \cdots + b_1(j\omega)U(j\omega) + b_0 U(j\omega),\end{aligned} \tag{10.6}$$

where $U(\omega) = \mathcal{F}[u(t)]$ and $Y(\omega) = \mathcal{F}[Y(t)]$. If we define

$$G(\omega) = \frac{Y(\omega)}{U(\omega)} = \frac{b_m(j\omega)^m + b_{m-1}(j\omega)^{m-1} + \cdots + b_1(j\omega) + b_0}{(j\omega)^n + a_{n-1}(j\omega)^{n-1} + \cdots + a_1(j\omega) + a_0}, \tag{10.7}$$

it describes the characteristics of the system without collecting the input and output signals because the coefficients a_i $(i = 0, 1, 2, \ldots, n)$ and b_j $(j = 0, 1, 2, \ldots, m)$ provide us with that information.

We use an example to show how to calculate the frequency response of a system.

[Example 10.1] Consider an ordinary differential equation

$$\frac{\mathrm{d}^2x(t)}{\mathrm{d}t^2} + 3\frac{\mathrm{d}x(t)}{\mathrm{d}t} + 2x(t) = r(t) \tag{10.8}$$

that describes the relationship between the input $[r(t)]$ and output $[x(t)]$. Assume that the initial conditions are $\mathrm{d}x(t)/\mathrm{d}t|_{t=0} = 0$ and $x(0) = 0$.

Performing the Fourier transform of (10.8)

$$\mathcal{F}\left[\frac{\mathrm{d}^2x(t)}{\mathrm{d}t^2} + 3\frac{\mathrm{d}x(t)}{\mathrm{d}t} + 2x(t)\right] = \mathcal{F}[r(t)]$$

gives

$$(j\omega)^2X(\omega) + 3(j\omega)X(\omega) + 2X(\omega) = R(\omega), \tag{10.9}$$

where $X(\omega)$ and $R(\omega)$ are the Fourier transforms of $x(t)$ and $r(t)$, respectively. If we define

$$G(\omega) = \frac{X(\omega)}{R(\omega)} = \frac{1}{2 - \omega^2 + 3\omega j}, \tag{10.10}$$

then we calculate its gain and phase

$$20\log_{10}|G(\omega)| = 20\log_{10}\frac{1}{\sqrt{(2-\omega^2)^2 + (3\omega)^2}}$$

$$= -20\log_{10}\sqrt{(2-\omega^2)^2+(3\omega)^2}, \tag{10.11}$$

$$\angle G(\omega) = -\tan^{-1}\frac{3\omega}{2-\omega^2} \tag{10.12}$$

and plot them separately to show the frequency response of the system (10.8) (**Figure 10.1**).

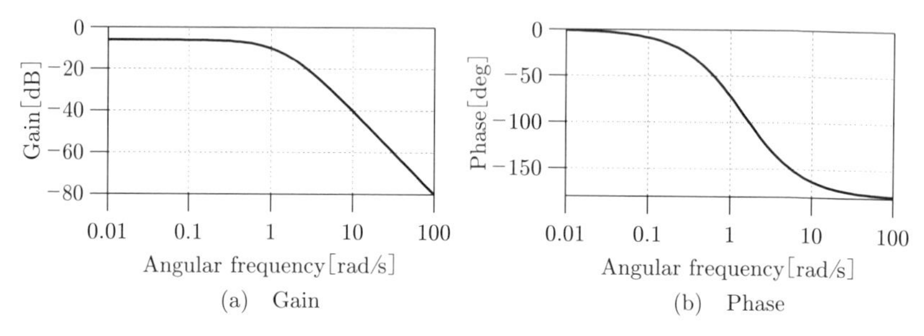

(a) Gain (b) Phase

Figure 10.1 Frequency response of system (10.8)

10.3 Comparison of Fourier and Laplace Transforms

Consider $r(t) = \sin t$ in (10.8). First, we use the Fourier transform to solve this equation. Then, we use the Laplace transform to solve it again and compare the solutions.

Since

$$R(\omega) = j\pi[\delta(\omega+1) - \delta(\omega-1)],$$

substituting it into (10.9) yields

$$X(j\omega) = j\pi\frac{\delta(\omega+1)-\delta(\omega-1)}{2-\omega^2+3\omega j}. \tag{10.13}$$

The solution $x(t)$ is given by the inverse Fourier transform of (10.13)

$$x(t) = \mathcal{F}^{-1}[X(j\omega)] = \frac{1}{2\pi}\int_{-\infty}^{\infty} X(j\omega)\mathrm{e}^{j\omega t}\mathrm{d}\omega$$

$$
\begin{aligned}
&= \frac{1}{2\pi}\int_{-\infty}^{\infty} j\pi \frac{\delta(\omega+1)-\delta(\omega-1)}{2-\omega^2+3\omega j} \mathrm{e}^{j\omega t}\mathrm{d}\omega \\
&= \frac{j}{2}\left[\int_{-\infty}^{\infty} \frac{\delta(\omega+1)}{2-\omega^2+3\omega j}\mathrm{e}^{j\omega t}\mathrm{d}\omega - \int_{-\infty}^{\infty} \frac{\delta(\omega-1)}{2-\omega^2+3\omega j}\mathrm{e}^{j\omega t}\mathrm{d}\omega\right] \\
&= \frac{j}{2}\left[\int_{-\infty}^{\infty} \frac{\mathrm{e}^{j\omega t}}{2-\omega^2+3\omega j}\delta(\omega+1)\mathrm{d}\omega \right.\\
&\quad \left. - \int_{-\infty}^{\infty} \frac{\mathrm{e}^{j\omega t}}{2-\omega^2+3\omega j}\delta(\omega-1)\mathrm{d}\omega\right]. \qquad (10.14)
\end{aligned}
$$

Since

$$
\int_{-\infty}^{\infty} f(t)\delta(t-a)\mathrm{d}t = f(a) \qquad (10.15)
$$

holds for an arbitrary function $f(t)$, (10.14) becomes

$$
\begin{aligned}
&\frac{j}{2}\left[\int_{\infty}^{-\infty} \frac{\mathrm{e}^{j\omega t}}{2-\omega^2+3\omega j}\delta(\omega+1)\mathrm{d}\omega \right.\\
&\quad \left. - \int_{-\infty}^{\infty} \frac{\mathrm{e}^{j\omega t}}{2-\omega^2+3\omega j}\delta(\omega-1)\mathrm{d}\omega\right] \\
&= \frac{j}{2}\left[\frac{\mathrm{e}^{-jt}}{1-3j} - \frac{\mathrm{e}^{jt}}{1+3j}\right] \\
&= \frac{1}{10}\sin t - \frac{3}{10}\cos t = \frac{1}{\sqrt{10}}\sin(t+\phi), \qquad (10.16)
\end{aligned}
$$

where

$$
\phi = -\tan^{-1} 3. \qquad (10.17)
$$

Now, we use the Laplace transform to solve (10.8). Performing the Laplace transform of (10.8)

$$
\mathcal{L}\left[\frac{\mathrm{d}^2x(t)}{\mathrm{d}t^2} + 3\frac{\mathrm{d}x(t)}{\mathrm{d}t} + 2x(t)\right] = \mathcal{L}(\sin t).
$$

Thus,

$$
(s^2+3s+2)X(s) = \frac{1}{s^2+1}.
$$

As a result,

$$X(s) = \frac{1}{s^2+3s+2} \times \frac{1}{s^2+1}. \tag{10.18}$$

Decomposing (10.18) gives

$$X(s) = -\frac{1}{5}\frac{1}{s+2} + \frac{1}{2}\frac{1}{s+3} + \frac{1}{10}\frac{1}{s^2+1} - \frac{3}{10}\frac{s}{s^2+1}. \tag{10.19}$$

Inverse Laplace transform of (10.19) yields

$$\mathcal{L}^{-1}[X(s)] = \underbrace{-\frac{1}{5}\mathrm{e}^{-2t} + \frac{1}{2}\mathrm{e}^{-3t}}_{\text{Transient response}} + \underbrace{\frac{1}{10}\sin t - \frac{3}{10}\cos t}_{\text{Steady-state response}}. \tag{10.20}$$

A comparison between (10.16) and (10.20) shows that the Fourier transfer provides us only with the steady-state response, while the solution of the Laplace transform contains both the transient and steady-state responses. It is worth mentioning that† Figure 10.1 shows the gains of a steady-state response for sinusoidal waves with different frequencies, for example, the gain is $-10\,\mathrm{dB}$ $(= 1/\sqrt{10})$ at $\omega = 1\,\mathrm{rad/s}$, which is exactly given in (10.16).

Problems

Basic Level

【1】 Find the Laplace transforms of the following functions:

A. $\mathcal{L}(1)$. B. $\mathcal{L}(\cos at)$ (a: constant). C. $\mathcal{L}(\sin at)$ (a: constant).

D. $\mathcal{L}[\ddot{y}(t)]$ if $y(0) = 0$, $\dot{y}(t)|_{t=0} = 0$, and $Y(s) = \mathcal{L}[y(t)]$.

【2】 Find the inverse Laplace transforms of the following functions:

A. $\dfrac{1}{s(s+1)}$. B. $\dfrac{1}{s^2+s+1}$. C. $\dfrac{s+1}{s^2+5s+6}$.

D. $\dfrac{1}{s-a}$ (a: constant). E. $\dfrac{s}{s^2+9}$.

【3】 Use the Laplace transform to find the solutions to the following differential equations:

A. $2\dfrac{\mathrm{d}y}{\mathrm{d}t} + y = 1,\ y(0) = 0$.

---

† It is worth mentioning that ～：～これは役に立つ［重要な］ことですが。

B. $\dfrac{\mathrm{d}^2 y}{\mathrm{d}t^2} + 4\dfrac{\mathrm{d}y}{\mathrm{d}t} + 5y = 0,\ y(0) = 1,\ \left.\dfrac{\mathrm{d}y}{\mathrm{d}t}\right|_{t=0} = 0.$

Advanced Level

【1】 Find the Laplace transforms of the following functions:

A. $\displaystyle\int_0^t \sin(t-x)\cos x \mathrm{d}x,$ B. $\displaystyle\int_0^t (1-x)\mathrm{e}^{a(t-x)}\mathrm{d}x.$

【2】 Using the Laplace transform to solve the following differential equations and compare the results:

A. $\ddot{y}(t) + 4y(t) = 1,$ B. $\ddot{y}(t) + 5\dot{y}(t) + 4y(t) = 1.$

【3】 Use the Laplace transform to find a unique solution for one-dimensional wave equation $\dfrac{\partial^2 u(x,t)}{\partial t^2} = c^2 \dfrac{\partial^2 u(x,t)}{\partial x^2}$ for the boundary conditions $u(0,t) = f(t)$ and $\lim\limits_{x\to\infty} u(x,t) = 0$, and the initial conditions $u(x,0) = 0$ and $\left.\dfrac{\partial u(x,t)}{\partial t}\right|_{t=0} = 0.$

Appendix

Table 1 Fourier transforms of typical functions

$f(t)$	$F(\omega)$
$\begin{cases} E, & \text{if } \lvert t\rvert \leqq \tau/2, \\ 0, & \text{otherwise} \end{cases}$	$2E\dfrac{\sin\dfrac{\omega\tau}{2}}{\omega}$
$\begin{cases} \dfrac{2A}{\tau}\left(\dfrac{\tau}{2}+t\right), & \text{if } -\dfrac{\tau}{2}\leqq t<0, \\ \dfrac{2A}{\tau}\left(\dfrac{\tau}{2}-t\right), & \text{if } 0\leqq t<\dfrac{\tau}{2} \end{cases}$	$\dfrac{4A}{\tau\omega^2}\left(1-\cos\dfrac{\omega\tau}{2}\right)$
$\begin{cases} E\cos\omega_0 t, & \text{if } \lvert t\rvert \leqq \dfrac{\tau}{2}, \\ 0, & \text{otherwise} \end{cases}$	$\dfrac{E\tau}{2}\left[\dfrac{\sin(\omega-\omega_0)\dfrac{\tau}{2}}{(\omega-\omega_0)\dfrac{\tau}{2}}+\dfrac{\sin(\omega+\omega_0)\dfrac{\tau}{2}}{(\omega+\omega_0)\dfrac{\tau}{2}}\right]$
$\mathrm{e}^{-\beta t}1(t)\ (\beta>0)$	$\dfrac{1}{\beta+j\omega}$
$\mathrm{e}^{-\beta\lvert t\rvert}\ (\beta>0)$	$\dfrac{2\beta}{\beta^2+\omega^2}$
$t\mathrm{e}^{-\beta t}1(t)\ (\beta>0)$	$\dfrac{1}{(\beta+j\omega)^2}$
$\dfrac{t^{n-1}}{(n-1)!}\mathrm{e}^{-\beta t}1(t)\ (\beta>0)$	$\dfrac{1}{(\beta+j\omega)^n}$

Table 1 (continued) Fourier transforms of typical functions

$f(t)$	$F(\omega)$
$Ae^{-\beta t^2}$ $(\beta > 0)$	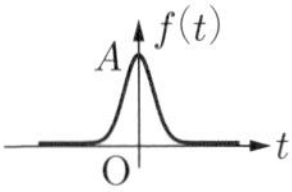 $\sqrt{\frac{\pi}{\beta}}Ae^{-\frac{\omega^2}{4\beta}}$ 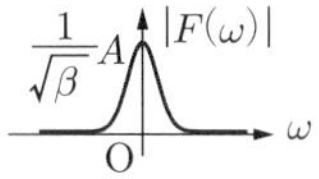
$\frac{1}{\sqrt{2\pi}\sigma}e^{-\frac{t^2}{2\sigma^2}}$	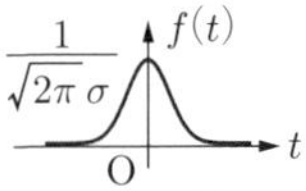 $e^{\frac{-\sigma^2\omega^2}{2}}$ 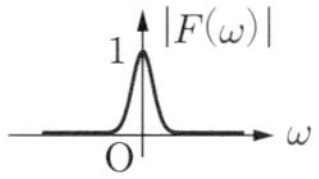
$\frac{\sin\omega_0 t}{\pi t}$	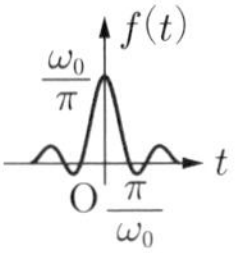 $\begin{cases} 1, & \text{if } \lvert\omega\rvert \leq \omega_0, \\ 0, & \text{otherwise} \end{cases}$ 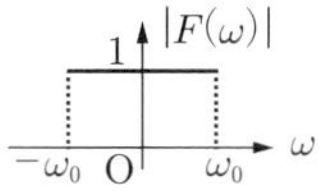
$\delta(t)$	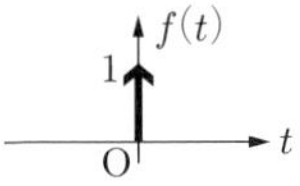 1 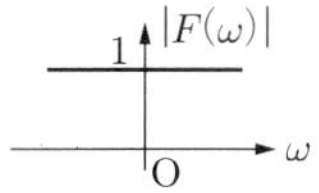
$1(t)$	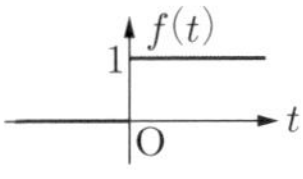 $\frac{1}{j\omega} + \pi\delta(t)$ 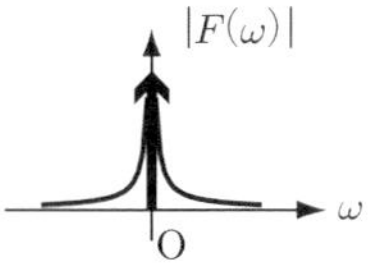
$\sum_{n=-\infty}^{\infty} \delta(t - nT)$	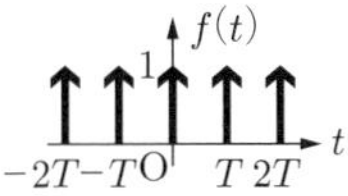 $\frac{2\pi}{T}\sum_{n=-\infty}^{\infty} \delta\left(\omega - \frac{2n\pi}{T}\right)$ 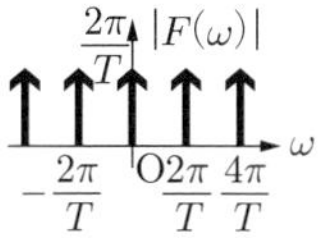
$\sin\omega_0 t$	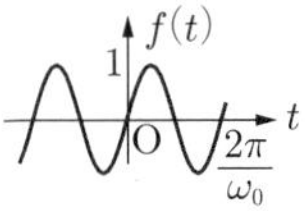 $j\pi[\delta(\omega+\omega_0)+\delta(\omega-\omega_0)]$ 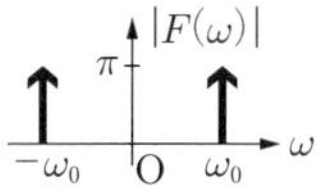
$\cos\omega_0 t$	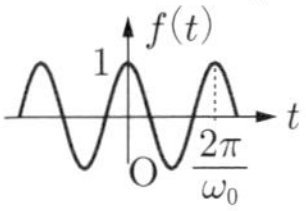 $\pi[\delta(\omega+\omega_0)+\delta(\omega-\omega_0)]$ 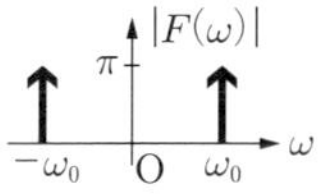
$e^{-\beta t}\sin\omega_0 t \times 1(t)$	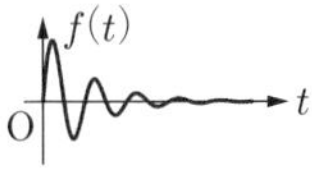 $\frac{\omega_0}{(j\omega+\beta)^2+\omega_0^2}$ 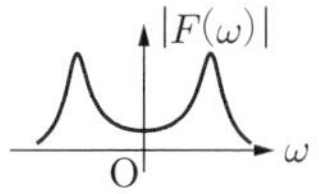
$e^{-\beta t}\cos\omega_0 t \times 1(t)$	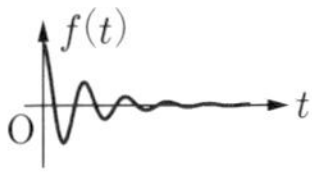 $\frac{j\omega+\beta}{(j\omega+\beta)^2+\omega_0^2}$ 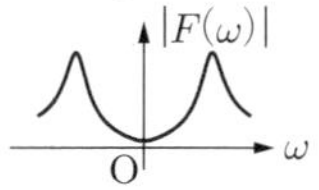

Table 2 Properties of Fourier transform

$f(t)$	$F(\omega)$	$f(t)$	$F(\omega)$
$a_1 f_1(t) + a_2 f_2(t)$	$a_1 F_1(\omega) + a_2 F_2(\omega)$	$f(t)\cos\omega_0 t$	$\frac{1}{2}\left[F(\omega-\omega_0)+F(\omega+\omega_0)\right]$
$f(t) = f_e(t) + f_o(t)$	$F(\omega) = X(\omega) + jY(\omega)$	$f(t)\sin\omega_0 t$	$\frac{1}{2j}\left[F(\omega-\omega_0)-F(\omega+\omega_0)\right]$
$f_e(t) = \dfrac{f(t)+f(-t)}{2}$	$X(\omega)$	$\dfrac{\mathrm{d}f(t)}{\mathrm{d}t}$	$j\omega F(\omega)$
$f_e(t) = \dfrac{f(t)-f(-t)}{2}$	$jY(\omega)$	$\dfrac{\mathrm{d}^n f(t)}{\mathrm{d}t^n}$	$(j\omega)^n F(\omega)$
$F(t)$	$2\pi f(-\omega)$	$\int_{-\infty}^{t} f(\tau)\mathrm{d}\tau$	$\dfrac{1}{j\omega} + \pi F(0)\delta(\omega)$
$f(at)$	$\dfrac{1}{\lvert a\rvert}F\left(\dfrac{\omega}{a}\right)$	$-jtf(t)$	$\dfrac{\mathrm{d}F(\omega)}{\mathrm{d}\omega}$
$f(-t)$	$F(-\omega)$	$(-jt)^n f(t)$	$\dfrac{\mathrm{d}^n F(\omega)}{\mathrm{d}\omega^n}$
$f(t-t_0)$	$F(\omega)\mathrm{e}^{-j\omega t_0}$	$\int_{-\infty}^{\infty} f(\tau)g(t-\tau)\mathrm{d}\tau$	$F(\omega)G(\omega)$
$f(t)\mathrm{e}^{-j\omega_0 t}$	$F(\omega-\omega_0)$	$f(t)g(t)$	$\dfrac{1}{2\pi}\int_{-\infty}^{\infty} F(\tau)G(\omega-\tau)\mathrm{d}\tau$

Table 3 Laplace transforms of typical functions

$f(t)$	$F(s)$	$f(t)$	$F(s)$
$\delta(t)$	1	$\sqrt{t}$	$\dfrac{\sqrt{\pi}}{2s^{3/2}}$
$1(t)$	$\dfrac{1}{s}$	$\dfrac{1}{\sqrt{t}}$	$\sqrt{\dfrac{\pi}{s}}$
t	$\dfrac{1}{s^2}$	$t^n\mathrm{e}^{-at}$	$\dfrac{n!}{(s+a)^{n+1}}$
t^n	$\dfrac{n!}{s^{n+1}}$	$t\sin\omega t$	$\dfrac{2\omega s}{(s^2+\omega^2)^2}$
e^{-at}	$\dfrac{1}{s+a}$	$t\cos\omega t$	$\dfrac{s^2-\omega^2}{(s^2+\omega^2)^2}$
$\sin\omega t$	$\dfrac{\omega}{s^2+\omega^2}$	$\mathrm{e}^{-at}\sin\omega t$	$\dfrac{\omega}{(s+a)^2+\omega^2}$
$\cos\omega t$	$\dfrac{s}{s^2+\omega^2}$	$\mathrm{e}^{-at}\cos\omega t$	$\dfrac{s+a}{(s+a)^2+\omega^2}$
$1(t-a)=\begin{cases} 0, & \text{if } t<a, \\ 1, & \text{if } t>a \end{cases}$	$\dfrac{\mathrm{e}^{-as}}{s}\ (a>0)$	$1(t-a)f(t-a)$	$\mathrm{e}^{-as}F(s)\ (a>0)$

References

1) N. M. Abbasi: The application of Fourier analysis in solving the Computed Tomography (CT) inverse problem, [online] Available: https://www.12000.org/my_notes/EE518_CT_project/REPORT/index.htm (April 22, 2022)

2) K. J. Åström and B. Wittenmark: Computer-Controlled Systems—Theory and Design—, 3rd Ed., Prentice Hall (1997)

3) 馬場 敬之：フーリエ解析 – キャンパス・ゼミ— 改訂 4, マセマ出版社 (2017)

4) S. L. Brunton and J. N. Kutz: Data-driven science and engineering: Machine learning, dynamical systems, and control, Cambridge University Press (2019)

5) S. L. Campbell, J.-P. Chancelier, and R. Nikoukhah: Modeling and Simulation in Scilab/Scicos with ScicosLab 4.4, 2nd Ed., Springer (2009)

6) J. W. Cooley and J. W. Tukey: An Algorithm for the Machine Calculation of Complex Fourier Series, Math. Computation, **19**, pp. 297–301 (1965)

7) Facts and details: Typhoons in Japan, [online] Available: https://factsanddetails.com/japan/cat26/sub160/item856.html (April 22, 2022)

8) 福田 安蔵, 鈴木 七緒, 安岡 善則, 黒崎 千代子：詳解応用解析演習，共立出版 (1970)

9) J. Fourier: The Analytical Theory of Heat, Cosimo Inc, Unabridged Ed. (2007)

10) 船越 満明：キーポイント フーリエ解析，岩波書店 (2006)

11) A. Gilat: MATLAB: An Introduction with Applications, 6th Ed., Wiley (2016)

12) 樋口 禎一, 森田 康夫：高校数学解法事典, 旺文社 (2003)

13) 石村 園子：やさしく学べるラプラス変換・フーリエ解析, 共立出版 (2015)

14) J. F. James: A Student's Guide to Fourier Transforms With Applications in Physics and Engineering, 3rd Ed., Cambridge University Press (2011)

15) H. Jung: Basic Physical Principles and Clinical Applications of Computed Tomography, Progress in Medical Physics, **32**, 1, online available: https://doi.org/10.14316/pmp.2021.32.1.1 (2021)

16) T. Kojima, H. Tsunashima, A. Matsumoto, and T. Mizuma: Anomaly Diagnosis of Railway by Service Train, [online] Available: http://www.me.cit.nihon-u.ac.jp/lab/tsuna/events/kojima.pdf (April 22, 2022)
17) G. G. Kuma, S. K. Sahoo, and P. K. Meher: 50 Years of FFT Algorithms and Applications, Circuits, Systems, and Signal Processing, **28**, pp. 5665–5698 (2019)
18) 楠田 信, 平居 孝之, 福田 亮治：使える数学 フーリエ・ラプラス変換, 共立出版 (1997)
19) J. N. Kutz, S. L. Brunton, B. W. Brunton, and J. L. Proctor: Dynamic Mode Decomposition—Data-Driven Modeling of Complex Systems, SIAM (2016)
20) 南京工学院数学教研組：工程数学　積分変換, 人民教育出版社 (1979)
21) K. Miyamoto, J. She, J. Imani, X. Xin, and D. Sato: Equivalent-input-disturbance approach to active structural control for seismically excited buildings, Engineering Structures, **125**, pp. 392–399 (2016)
22) K. Miyamoto, S. Nakano, J. She, D. Sato, Y. Chen, and Q.-L. Han: Design Method of Tuned Mass Damper by Linear-Matrix-Inequality-Based Robust Control Theory for Seismic Excitation, Journal of Vibration and Acoustics, **144**, 4, pp. 041008: 1 $\sim$ 14 (2022)
23) 森下 巖, 小畑 秀文：信号処理, コロナ社 (1982)
24) National Institute for Occupational Safety and Health, Selected Topics in Surface Electromyography for Use in the Occupational Setting: Expert Perspective, U.S. Department of Health and Human Services (1992)
25) N. Morrison: Introduction to Fourier Analysis, John Wiley & Sons (1995)
26) 大石 進一：フーリエ解析, 岩波書店 (1995)
27) 大崎 順彦：新・地震動のスペクトル解析入門, 鹿島出版会 (2011)
28) S. J. Orfanidis: Introduction to Signal Processing, Prentice Hall (1995)
29) W. H. Press and W. T. Vetterling: Numerical Recipes in C, Cambridge University Press (1997)
30) J. G. Proakis and D. G. Manolakis: Digital Signal Processing: Principles, Algorithms, and Applications, Prentice Hall (1996)
31) J. She, Y. Ohyama, and H. Kobayashi: Master-Slave Electric Cart Control System for Maintaining/Improving Physical Strength, IEEE Trans.

Robotics, **22**, 3, pp. 481–490 (2006)

32) 末松 良一, 山田 宏尚：画像処理工学（改訂版）, コロナ社 (2016)

33) The Ministry of Land, Infrastructure, Transport and Tourism: Find Out! Japan's Preparations for Earthquakes, [online] Available: https://www.mlit.go.jp/river/earthquake/en/index.html (April 22, 2022)

34) 田中 敏幸：X 線 CT の原理・現状とさらなる画像の高品質化, 計測と制御, **56**, 11, pp. 874–879 (2017)

35) 塚原 直樹：カラスをだます, NHK 出版 (2021)

36) 塚原 直樹, 永田 健：カラスの音声コミュニケーションとそれを応用した被害対策, 日本音響学会 騒音・振動研究会資料, N-2022-07, pp. 1–7 (2022)

37) A. Vande Wouwer, P. Saucez, and C. Vilas Fernández: Simulation of ODE/PDE Models with MATLAB®, OCTAVE and SCILAB, Springer (2014)

38) 涌井 良幸, 涌井 真美：道具としてのフーリエ解析, 日本実業出版 (2014)

39) 和達 三樹：物理のための数学, 岩波書店 (1983)

40) W. Y. Yang, Y. S. Cho, C. W. Choo, W. G. Jeon, J. Kim, S. Yu, K. W. Park, J. Wee, and K. H. Prashantha: MATLAB/Simulink for Digital Signal Processing, Hongrung Publishing Company (2014)

41) J. Zhao, J. She, E. F. Fukushima, D. Wang, Min Wu, and K. Pan: Muscle Fatigue Analysis With Optimized Complementary Ensemble Empirical Mode Decomposition and Multi-Scale Envelope Spectral Entropy, Frontiers in Neurorobotics, **14**, pp. 566172:1–14, DOI: 10.3389/fnbot.2020.566172 (2020)

Answers to Problems

★ Chapter 1

Basic Level

【1】【2】 Omitted.

Advanced Level

【1】 (1) Split an image into blocks of 8×8 pixels. (2) Carry out a discrete cosine transform (a kind of Fourier transform) for each block to convert each block from the spatial (2-D) domain into the frequency domain. (3) Quantize the amplitudes of the frequency components to compress the data. (4) A lossless algorithm further compresses the resulting data.

【2】 (1) Take X-ray images from different angles around a body. (2) Carry out a Fourier transform for each of the X-ray images. (3) Carry out an inverse Fourier transform to create cross-sectional images (slices) of the bones, blood vessels, and soft tissues inside the body.

★ Chapter 2

Basic Level

【1】 See **Table A2.1**.

Table A2.1 Solution of Basic-Level Problem 1

No.	Cartesian coordinate	Polar coordinate	Conjugate number
A	$(\sqrt{2}, \sqrt{2})$	$2\mathrm{e}^{j\pi/4}$	$\sqrt{2} - j\sqrt{2}$
B	$\left(\frac{1}{4}, -\frac{1}{4}\right)$	$\frac{\sqrt{2}}{4}\mathrm{e}^{-j\pi/4}$	$\frac{1}{4} + j\frac{1}{4}$
C	$\left(\frac{3}{2}, -\frac{5}{2}\right)$	$\frac{\sqrt{34}}{2}\mathrm{e}^{j\theta}\ \left(\theta = -\tan^{-1}\frac{5}{3}\right)$	$\frac{3}{2} + j\frac{5}{2}$
D	$(0, -4)$	$4\mathrm{e}^{-j\pi/2}$	$j4$

[2] A. $a+b=\dfrac{1+j\sqrt{3}}{2}+\dfrac{1-j\sqrt{3}}{2}=1$ and $ab=\dfrac{1+j\sqrt{3}}{2}\times\dfrac{1-j\sqrt{3}}{2}=1$. Thus, $a^3+b^3-3ab=(a+b)^3-3ab(a+b)-3ab=1^3-3\times1\times1-3\times1=-5$.

B. $x=\dfrac{1+j\sqrt{3}}{2}$ provides $2x-1=j\sqrt{3}$. Squaring both sides gives $4x^2-4x+1=-3$, that is, $x^2-x+1=0$. Dividing x^4-3x^2+6x-4 by x^2-x+1 obtains the quotient x^2+x-3 and the remainder $2x-1$. Thus, $x^4-3x^2+6x-4=(x^2-x+1)(x^2+x-3)+2x-1=2x-1=2\times\dfrac{1+j\sqrt{3}}{2}-1=j\sqrt{3}$.

[3] $x=1$ and $y=11$.

[4] A. $-1=\cos\pi+j\sin\pi=\mathrm{e}^{j\pi}$.

B. $$1-\cos\phi+j\sin\phi=2\sin\frac{\phi}{2}\left[\cos\left(\frac{\pi}{2}-\frac{\phi}{2}\right)+j\sin\left(\frac{\pi}{2}-\frac{\phi}{2}\right)\right]$$
$$=2\sin\frac{\phi}{2}\mathrm{e}^{j(\pi/2-\phi/2)}.$$

C. $\dfrac{j2}{-1+j}=\sqrt{2}\left(\cos\dfrac{\pi}{4}-j\sin\dfrac{\pi}{4}\right)=\sqrt{2}\mathrm{e}^{-j\pi/4}$.

D. $\dfrac{(\cos5\phi+j\sin5\phi)^2}{(\cos3\phi-j\sin3\phi)^3}=\dfrac{\left(\mathrm{e}^{j5\phi}\right)^2}{\left(\mathrm{e}^{-j3\phi}\right)^3}=\mathrm{e}^{j19\phi}$.

[5] A. $\dfrac{1}{2}$. B. 0. C. 1. D. 1.

[6] A. $\cos x$. B. $-\sin x$. C. $\cos x+\dfrac{1}{x}$. D. $2\cos2x$.

E. $\sin x+x\cos x$. F. $(n+x)x^{n-1}\mathrm{e}^x$.

[7] A. $2xy$. B. $y^2-y^4\sin x$. C. $-2x\mathrm{e}^{-(x^2+y^2)}$.

[8] A. $-\cos x+C$. B. $\dfrac{1}{2}\mathrm{e}^x(\sin x-\cos x)+C$.

C. $x(\log_e x-1)\log_{10}\mathrm{e}+C$. D. $\mathrm{e}^{x+2}+C$.

[9] $f(2)=2^n\times\alpha+\beta=17\ \cdots\ (a1)$, $f(4)=4^n\times\alpha+\beta=77\ \cdots\ (a2)$, and $f(8)=8^n\times\alpha+\beta=377\ \cdots\ (a3)$. $(a2)-(a1)$: $(4^n-2^n)\alpha=60\ \cdots\ (a4)$. $\therefore\ 2^n(2^n-1)\alpha=60$. $(a3)-(a2)$: $(8^n-4^n)\alpha=300$. $\therefore\ 4^n(2^n-1)\alpha=300\ \cdots\ (a5)$. $(a5)/(a4)$: $2^n=5$. Substituting it into $(a4)$ gives $5\times4\alpha=60$. $\therefore\ \alpha=3$. Frome $(a1)$, we have $5\times3+\beta=17$. $\therefore\ \beta=2$.

[10] Let $x=\log_{10}2$. Hence, $10^x=2$. Note $10^{0.3}=\sqrt[10]{10^3}=\sqrt[10]{1000}$. Since $2^{10}=1024$, $10^{0.3}=\sqrt[10]{1000}<\sqrt[10]{1024}=2=10^x$. Thus, $10^{0.3}<10^x$ and $0.3<x$. In the same way, $10^{0.4}=\sqrt[10]{10^4}=\sqrt[10]{10000}$. $10^{0.4}=\sqrt[10]{10000}>\sqrt[10]{1024}=2=10^x$. Thus, $0.3<x<0.4$.

【11】 A. $\dfrac{\pi}{6}$ rad. B. $\dfrac{5\pi}{6}$ rad. C. $\dfrac{5\pi}{4}$ rad. D. $114.59°$.
E. $180°$. F. $28.65°$.

【12】 A. $\dfrac{\sqrt{3}}{2}$. B. $\dfrac{\sqrt{3}}{3}$. C. $\dfrac{\sqrt{2}}{2}$. D. $-\dfrac{\sqrt{3}}{6}$.

【13】 Let the period be T. Thus, $\cos\left(\dfrac{t+T}{2}\right)+\cos\left(\dfrac{t+T}{7}\right)=\cos\dfrac{t}{2}+\cos\dfrac{t}{7}$.
Since $\cos(\phi+2k\pi)=\cos\phi$ holds for an integer k, $\dfrac{T}{2}=2m\pi$ and $\dfrac{T}{7}=2n\pi$ hold for integers m and n. As a result, $T=4m\pi=14n\pi$. It holds for $m=7$ and $n=2$. Hence, $T=28\pi$.

【14】 Eq. (2.43): $\displaystyle\int_{-\infty}^{\infty}\phi(t)1(t)\mathrm{d}t=\int_{-\infty}^{0}\phi(t)\times 0\mathrm{d}t+\int_{0}^{\infty}\phi(t)\times 1\mathrm{d}t=\int_{0}^{\infty}\phi(t)\mathrm{d}t.$
Eq. (2.44): Let $\tau=-t$. $\delta(\tau)=0$ if $\tau\neq 0$ and $\delta(\tau)=\infty$ if $\tau=0$. Thus, $\delta(-t)=\delta(t)$ holds from the definition of $\delta(t)$,

Eq. (2.45): Letting $\tau=at$ and examining $a>0$ and $a<0$ give
$\displaystyle\int_{-\infty}^{\infty}\delta(at)\phi(t)\mathrm{d}t=\frac{1}{|a|}\int_{-\infty}^{\infty}\delta(\tau)\phi\left(\frac{\tau}{a}\right)\mathrm{d}\tau=\frac{1}{|a|}\phi(0).$

Eq. (2.46): Letting $\tau=t-t_0$ gives $\displaystyle\int_{-\infty}^{\infty}\phi(t)\delta(t-t_0)\mathrm{d}t$

$$=\int_{-\infty}^{\infty}\phi(\tau+t_0)\delta(\tau)\mathrm{d}\tau$$
$$=\int_{-\infty}^{\infty}\phi(\tau+t_0)\left[\lim_{\epsilon\to 0}\delta_\epsilon(\tau)\right]\mathrm{d}\tau=\lim_{\epsilon\to 0}\int_{-\infty}^{\infty}\phi(\tau+t_0)\delta_\epsilon(\tau)\mathrm{d}\tau$$
$$=\lim_{\epsilon\to 0}\int_{0}^{\epsilon}\phi(\tau+t_0)\frac{1}{\epsilon}\mathrm{d}\tau=\lim_{\epsilon\to 0}\phi(\theta)\ (t_0\leqq\theta\leqq t_0+\epsilon)$$
(From the integral mean value theorem) $=\phi(t_0)$.

Eq. (2.47): $\displaystyle\int_{-\infty}^{\infty}\delta(at)\phi(t)\mathrm{d}t=\frac{1}{|a|}\phi(0)=\frac{1}{|a|}\int_{-\infty}^{\infty}\delta(t)\phi(t)\mathrm{d}t$
$\displaystyle=\int_{-\infty}^{\infty}\frac{1}{|a|}\delta(t)\phi(t)\mathrm{d}t$. Hence, $\delta(at)=\dfrac{1}{|a|}\delta(t)$.
Eq. (2.48): Letting $a=-1$ in (2.47) yields the result.

Advanced Level

【1】 A point $P(x,y)$ that is given by $x=\cos\theta$ and $y=\sin\theta$ $(0\leqq\theta<2\pi)$ is a circle: $x^2+y^2=1$ (**Figure A2.1**). Choose a point A to be $(2,2)$. Then, $f(\theta)$ is the gradient of the straight line AP. It has the minimum and maximum when AP, $y-2=m(x-2)$, is the tangent line of the circle.

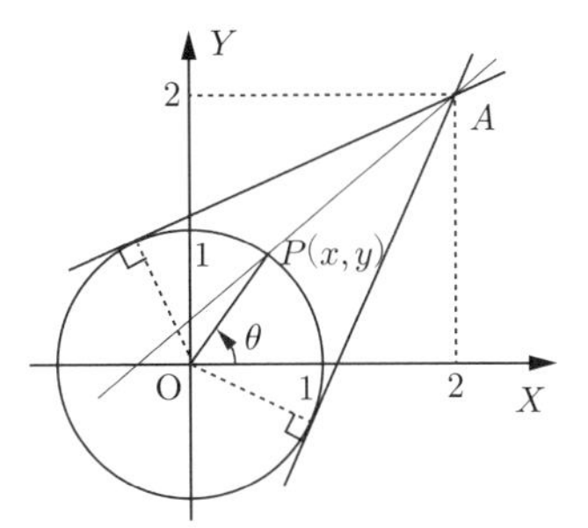

Figure A2.1 Advanced-Level Problem 1

This holds when the length of the perpendicular to the straight line AP from the origin is 1, that is, $\dfrac{|2m-2|}{\sqrt{1+m^2}} = 1$. $\therefore$ $4(m-1)^2 = m^2+1$, $3m^2-8m+3=0$. Thus, $m = \dfrac{4\pm\sqrt{7}}{3}$. As a result, maximum: $\dfrac{4+\sqrt{7}}{3}$ and minimum: $\dfrac{4-\sqrt{7}}{3}$.

[2] A. Taking a power of y on both sides of $m = a^x$ by y yields $m^y = a^{xy}$, and taking a power of x on both sides of $n = a^y$ by x yields $n^x = a^{xy}$. Applying the third condition to these two new equations yields $a^{xy} \times a^{xy} = a^{2/z}$. $\therefore$ $a^{2xy} = a^{2/z}$. Since $a \neq 1$, $2xy = \dfrac{2}{z}$. $\therefore$ $xyz = 1$.

B. The condition is $2^{3x} = 3^{2y} = 2^z \cdot 3^z$. Thus, $\left(2^{3x}\right)^{z/(2y)} = \left(3^{2y}\right)^{z/(2y)}$, that is, $2^{3xz/(2y)} = 3^z$. Substituting it into $2^{3x} = 2^z \cdot 3^z$ yields $2^{3x} = 2^{z+3xz/(2y)}$. Thus, $3x = z + \dfrac{3xz}{2y}$, that is, $\dfrac{2}{x} + \dfrac{3}{y} = \dfrac{6}{z}$.

[3] A. Since $\log_y x = \dfrac{1}{\log_x y}$, $\log_x y = 2 + \dfrac{3}{\log_x y}$. $(\log_x y)^2 - 2\log_x y - 3 = 0$. $\log_x y = 3$ and $\log_x y = -1$.

B. Form A, $y = x^3$ and $y = \dfrac{1}{x}$ for $x > 0$ and $x \neq 1$.

[4] Note that the inequality holds for $a \leqq 0$ and $x > 0$. Taking the logarithm of the base 2 of both sides for $a > 0$ yields $(\log_2 x)^2 \geqq 2\log_2 x + \log_2 a$. Letting $X = \log_2 x$ gives $X^2 - 2X - \log_2 a \geqq 0$ $\cdots$ $(a1)$. When $x > 0$, X is a real number. Thus, $(a1)$ holds if the discriminant of $(a1) \leqq 0$, that is, $1 + \log_2 a \leqq 0$, or $\log_2 a \leqq 0 \leqq -1$. $\therefore$ $a \leqq \dfrac{1}{2}$.

[5] Let $f(t)$ be a continuous function. $\displaystyle\int_{-\infty}^{\infty} t\delta(t)\mathrm{d}t = \lim_{\epsilon\to\infty}\int_{-\epsilon}^{\epsilon} t\delta_\epsilon(t)\mathrm{d}t =$

$\lim_{\epsilon\to 0} 2\epsilon\zeta\delta_\epsilon(\zeta)$ (Mean value theorem for integrals), where $-\epsilon \leqq \zeta \leqq \epsilon$. On the other hand, $\int_{-\infty}^{\infty} t\delta(t)\mathrm{d}t = 0$. The equation $\lim_{\epsilon\to 0}\epsilon[\zeta\delta_\epsilon(\zeta)] = 0$ implies that $\lim_{\epsilon\to 0}\zeta\delta_\epsilon(\zeta)$ must be finite. The continuity of $\zeta\delta_\epsilon(\zeta)$ in $[-\epsilon,\epsilon]$ provides us with $t\delta(t) = 0$.

★ Chapter 3

Basic Level

【1】 A. Odd. B. None. C. Even. D. Even. E. Even. F. Odd. G. Even. H. Odd if $b = 0$ and $n = 1, 3, 5, \ldots$; even if $n = 2, 4, 6, \ldots$.

【2】 Suppose that $f(t)$ is periodic with period T. Thus, there must exist integers m and n such that $T = 2m\pi$ and $(1+\pi)T = 2n\pi$. Takin the ratio of both sides yields $1 + \pi = n/m$. Since π is an irrational number, the left side of this equation is an irrational number. However, the right side is a rational number due to the ratio of integers, Hence, this equation does not hold. As a result, $f(t) = \cos t + \cos(1+\pi)t$ is not periodic.

【3】 A. $\dfrac{2\pi}{n}$ $(n \neq 0)$. B. $\dfrac{1}{2}$. C. $2k$. D. π. E. 8π.

【4】 A. Not piecewise continuous. B. Continuous.
C. Continuous and piecewise smooth. D. Piecewise smooth.
E. Continuous and piecewise smooth. F. Piecewise continuous.

【5】 $f(t) = \dfrac{f(t)+f(-t)}{2} + \dfrac{f(t)-f(-t)}{2}$. Let $f_e(t) = \dfrac{f(t)+f(-t)}{2}$ and $f_o(t) = \dfrac{f(t)-f(-t)}{2}$. Then, $f_e(t)$ is even and $f_o(t)$ is odd. Hence, $f(t) = f_e(t) + f_o(t)$.

【6】 Since $f(-t) = -f(t)$, $|f(-t)| = |-f(t)| = |f(t)|$. Thus, $|f(t)|$ is even.

【7】 A. $a_0 = 0$. Even function. Thus, $b_n = 0$.

$$a_n = \frac{4}{n\pi}\sin\frac{n\pi}{2} = \begin{cases} (-1)^{\frac{n+3}{2}}\dfrac{4}{n\pi}, & n = 1, 3, 5, \ldots, \\ 0, & n = 2, 4, 6, \ldots. \end{cases}$$

B. Odd function. Thus, $a_n = 0$ $(n = 0, 1, 2, \ldots)$.

$$b_n = \frac{2}{n\pi}(1 - \cos n\pi) = \begin{cases} \dfrac{4}{n\pi}, & n = 1, 3, 5, \ldots, \\ 0, & n = 2, 4, 6, \ldots. \end{cases}$$

C. Even function. Thus, $b_n = 0$. $a_0 = \dfrac{2\sin\lambda\pi}{\lambda\pi}$. $a_n = (-1)^n \dfrac{2\lambda\sin\lambda\pi}{\pi(\lambda^2 - n^2)}$ $(n = 0, 1, 2, \ldots)$.

D. $a_0 = \dfrac{2A}{\pi}$, $a_n = \begin{cases} 0, & n = 1, 3, 5, \ldots, \\ -\dfrac{2A}{(n-1)(n+1)\pi}, & n = 2, 4, 6, \ldots. \end{cases}$, $b_1 = \dfrac{A}{2}$, and $b_n = 0$ $(n = 2, 3, \ldots)$.

E. $a_0 = 1$, $a_n = 0$, and $b_n = -\dfrac{1}{n\pi}$.

F. $a_0 = 0$. Even function. Thus, $b_n = 0$.

$$a_n = \frac{4}{n^2\pi^2}(1 - \cos n\pi) = \begin{cases} \dfrac{8}{n^2\pi^2}, & n = 1, 3, 5, \ldots, \\ 0, & n = 2, 4, 6, \ldots. \end{cases}$$

G. Odd function. Thus, $a_n = 0$ $(n = 0, 1, 2, \ldots)$.

$b_n = (-1)^{n+1}\dfrac{2}{n}$ $(n = 1, 2, \ldots)$.

H. Even function. Thus, $b_n = 0$. $a_0 = \dfrac{2\pi^2}{3}$.

$a_n = (-1)^n \dfrac{4}{n^2}$ $(n = 1, 2, \ldots)$.

I. Odd function. Thus, $a_n = 0$ $(n = 0, 1, 2, \ldots)$.

$b_n = (-1)^{n+1}\left(\dfrac{2\pi^2}{n} - \dfrac{12}{n^3}\right)$ $(n = 1, 2, 3, \ldots)$.

【8】 Omitted.

【9】 A. $c_n = \dfrac{2}{n\pi}\sin\dfrac{n\pi}{2} = (-1)^{\frac{n+3}{2}}\dfrac{2}{n\pi}$ and $|c_n| = \dfrac{2}{|n|\pi}$ $(n = \pm 1, \pm 3, \pm 5, \ldots)$.
$\phi_n = 0^\circ$ $(n = \ldots, -7, -3, 1, 5, \ldots)$ and $\phi_n = -180^\circ$ $(n = \ldots, -5, -1, 3, 7, \ldots)$.

B. $c_n = \dfrac{2}{n\pi}$ and $|c_n| = \dfrac{2}{|n|\pi}$ $(n = \pm 1, \pm 3, \pm 5, \ldots)$. $\phi_n = 0^\circ$ $(n = 1, 3, \ldots)$ and $\phi_n = -180^\circ$ $(n = -1, -3, \ldots)$.

C. $c_n = (-1)^n \dfrac{\lambda\sin\lambda\pi}{\pi(\lambda^2 - n^2)}$ and $|c_n| = \left|\dfrac{\lambda\sin\lambda\pi}{\pi(\lambda^2 - n^2)}\right|$ $(n = 0, \pm 1, \pm 2, \ldots)$,

$$\phi_n = \begin{cases} 0^\circ, & \left(\dfrac{\lambda \sin \lambda\pi}{\lambda^2 - n^2} < 0 \ \& \ n = \pm 1, \pm 3, \pm 5, \ldots\right) \text{ or} \\ & \left(\dfrac{\lambda \sin \lambda\pi}{\lambda^2 - n^2} \geqq 0 \ \& \ n = 0, \pm 2, \pm 4, \ldots\right), \\ -180^\circ, & \left(\dfrac{\lambda \sin \lambda\pi}{\lambda^2 - n^2} \geqq 0 \ \& \ n = \pm 1, \pm 3, \pm 5, \ldots\right) \text{ or} \\ & \left(\dfrac{\lambda \sin \lambda\pi}{\lambda^2 - n^2} < 0 \ \& \ n = 0, \pm 2, \pm 4, \ldots\right). \end{cases}$$

D. $c_{-1} = j\dfrac{A}{4}$, $c_1 = -j\dfrac{A}{4}$, and $c_n = -\dfrac{A}{\pi(n-1)(n+1)}$ $(n = 0, \pm 2, \pm 4, \pm 6, \ldots)$. $|c_{-1}| = \dfrac{A}{4}$, $|c_1| = \dfrac{A}{4}$, and $|c_n| = \dfrac{A}{\pi|(n-1)(n+1)|}$ $(n = 0, \pm 2, \pm 4, \pm 6, \ldots)$. $\phi_{-1} = 90^\circ$, $\phi_1 = -90^\circ$, $\phi_0 = 0^\circ$, and $\phi_n = -180^\circ$ $(n = \pm 2, \pm 4, \pm 6, \ldots)$.

E. $c_0 = \dfrac{1}{2}$ and $c_n = j\dfrac{1}{2n\pi}$. $|c_0| = \dfrac{1}{2}$ and $|c_n| = \dfrac{1}{2|n|\pi}$ $(n = \pm 1, \pm 2, \ldots)$. $\phi_0 = 0^\circ$, $\phi_n = 90^\circ$ $(n = 1, 2, \ldots)$, and $\phi_n = -90^\circ$ $(n = -1, -2, \ldots)$.

F. $c_n = \dfrac{4}{n^2\pi^2}$, $|c_n| = \dfrac{4}{n^2\pi^2}$, and $\phi_n = 0^\circ$ $(n = \pm 1, \pm 3, \pm 5, \ldots)$.

G. $c_n = j(-1)^n\dfrac{1}{n}$ and $|c_n| = \dfrac{1}{|n|}$ $(n = \pm 1, \pm 2, \ldots)$. $\phi_n = 90^\circ$ $(n = \ldots, -3, -1, 2, 4, \ldots)$ and $\phi_n = -90^\circ$ $(n = \ldots, -4, -2, 1, 3, \ldots)$.

H. $c_0 = \dfrac{\pi^2}{3}$ and $c_n = (-1)^n\dfrac{2}{n^2}$ $(n = \pm 1, \pm 2, \ldots)$. $|c_0| = \dfrac{\pi^2}{3}$ and $|c_n| = \dfrac{2}{n^2}$ $(n = \pm 1, \pm 2, \ldots)$. $\phi_0 = 0^\circ$, $\phi_n = -180^\circ$ $(n = \pm 1, \pm 3, \pm 5, \ldots)$, and $\phi_n = 0^\circ$ $(n = \pm 2, \pm 4, \pm 6, \ldots)$.

I. $c_n = j(-1)^n\left(\dfrac{\pi^2}{n} - \dfrac{6}{n^3}\right)$ and $|c_n| = \left|\dfrac{\pi^2}{n} - \dfrac{6}{n^3}\right|$ $(n = \pm 1, \pm 2, \ldots)$. $\phi_n = -90^\circ$ $(n = \ldots, -4, -2, 1, 3, \ldots)$ and $\phi_n = 90^\circ$ $(n = \ldots, -3, -1, 2, 4, \ldots)$.

Advanced Level

【1】 The Fourier series of $f_T(t) = \begin{cases} -1, & \text{if} \quad -\pi < t < 0, \\ 0, & \text{if} \quad t = 0, \pm\pi, \\ 1, & \text{if} \quad 0 < t < \pi \end{cases}$

is $f_T(t) = \dfrac{4}{\pi}\left(\sin t + \dfrac{1}{3}\sin 3t + \dfrac{1}{5}\sin 5t + \cdots\right)$. Letting $t = \dfrac{\pi}{2}$ gives $1 =$

$\frac{4}{\pi}\left(1-\frac{1}{3}+\frac{1}{5}+\cdots\right).$

【2】 A. For Basic-Level Problem **【7】** G., $t=\sum_{n=1}^{\infty}(-1)^{n+1}\frac{2}{n}\sin nt$. Note that $\mathrm{d}t/\mathrm{d}t=1$ and $\mathrm{d}\left(\sum_{n=1}^{\infty}(-1)^{n+1}\frac{2}{n}\sin nt\right)/\mathrm{d}t=\sum_{n=1}^{\infty}2(-1)^{n+1}\cos nt$. If $t=0$, the series on the right-hand side becomes $2(1-1+1-1+1-1+\cdots)$ and diverges. If $t=\pi$, the series on the right-hand side becomes $2(-1-1-1-\cdots)$ and also diverges. Thus, the derivatives on the left and right sides are not the same.

For Basic-Level Problem **【7】** H., $t^2=\frac{\pi^2}{3}+\sum_{n=1}^{\infty}(-1)^n\frac{4}{n^2}\cos nt$. On the other hand, $\frac{\mathrm{d}t^2}{\mathrm{d}t}=2t$ and $\frac{\mathrm{d}}{\mathrm{d}t}\left[\frac{\pi^2}{3}+\sum_{n=1}^{\infty}(-1)^n\frac{4}{n^2}\cos nt\right]=\sum_{n=1}^{\infty}(-1)^n(-1)\frac{4}{n}\sin nt=2\sum_{n=1}^{\infty}(-1)^{n+1}\frac{2}{n}\sin nt=2t$. They are the same.

B. Since $\mathrm{d}f_T(t)/\mathrm{d}t$ is piecewise continuous and differentiable, its Fourier series converges to it. Let

$$f_T(t)=\frac{a_0}{2}+\sum_{n=1}^{\infty}(a_n\cos n\omega t+b_n\sin n\omega t)$$

and

$$\frac{\mathrm{d}f_T(t)}{\mathrm{d}t}=\frac{\alpha_0}{2}+\sum_{n=1}^{\infty}(\alpha_n\cos n\omega t+\beta_n\sin n\omega t),$$

where $\alpha_0=\frac{2}{T}\int_{-T/2}^{T/2}\frac{\mathrm{d}f_T(t)}{\mathrm{d}t}\mathrm{d}t$, $\alpha_n=\frac{2}{T}\int_{-T/2}^{T/2}\frac{\mathrm{d}f_T(t)}{\mathrm{d}t}\cos n\omega t\mathrm{d}t$, and $\beta_n=\frac{2}{T}\int_{-T/2}^{T/2}\frac{\mathrm{d}f_T(t)}{\mathrm{d}t}\sin n\omega t\mathrm{d}t$. As a result, $\alpha_0=0$, $\alpha_n=n\omega b_n$, and $\beta_n=-n\omega a_n$. Thus, the differentiation theorem holds.

【3】 The differentiation theorem requires that $f_T(-T/2)=f_T(T/2)$ and $f_T(t+T)=f_T(t)$.

【4】 Construct a continuous periodic function $g(t)=\int_0^t f(\tau)\mathrm{d}\tau-\frac{1}{2}a_0t$. $\mathrm{d}g(t)/\mathrm{d}t$ is also continuous. Let the Fourier series expansion of $g(t)$ be $g(t)=$

$\dfrac{\alpha_0}{2} + \sum_{n=1}^{\infty}(\alpha_n \cos n\omega t + \beta_n \sin n\omega t)$. Calculate the coefficients α_n ($n = 0, 1, 2, \ldots$) and β_m ($m = 1, 2, \ldots$). Finally, calculating $\int_{t_1}^{t_2} f_T(t)\mathrm{d}t = F(t_2) - F(t_1) + \dfrac{1}{2}a_0(t_2 - t_1)$ yields the result.

【5】 (Hint) Calculate the Fourier transforms of periodic functions t, t^2, t^3, and t^4 for $-\pi < t < \pi$ and combine them.

★ Chapter 4

Basic Level

【1】 A. $\mathcal{F}[f(t)] = \mathcal{F}\left[\dfrac{A(t)(\mathrm{e}^{j\omega_0 t} + \mathrm{e}^{-j\omega_0 t})}{2}\right] = \dfrac{\sin \dfrac{D(\omega - \omega_0)}{2}}{\omega - \omega_0} + \dfrac{\sin \dfrac{D(\omega + \omega_0)}{2}}{\omega + \omega_0}$.

B. $F(\omega) = \dfrac{4 \sin \omega}{\omega^3}\left(\dfrac{1}{\omega} + j\right)$. C. $F(\omega) = \dfrac{2}{5 - \omega^2 + j2\omega}$.

D. $F(\omega) = -\dfrac{4}{j\omega}\sin^2 \dfrac{\omega}{2}$. E. $F(\omega) = \cos a\omega + \cos \dfrac{a\omega}{2}$.

【2】 A. $F(\omega) = \dfrac{4A}{\tau\omega^2}\left(1 - \cos \dfrac{\tau\omega}{2}\right)$. B. $F(\omega) = \dfrac{\sigma^2\omega^2}{2}$.

【3】 (Hint) $F(\omega) = \int_{-\infty}^{\infty}\left(\sum_{n=-\infty}^{\infty} c_n \mathrm{e}^{jn\omega_0 t}\right)\mathrm{e}^{-j\omega t}\mathrm{d}t$.

【4】 A. $\dfrac{2a}{a^2 + \omega^2}$ and perform its inverse Fourier transform. B. $\dfrac{2(2 + \omega^2)}{4 + \omega^4}$ and perform its inverse Fourier transform. C. $-\dfrac{j2 \sin \omega\pi}{1 - \omega^2}$ and perform its inverse Fourier transform.

【5】 (Hint) Define a new variable $\tau = at$ and consider $a > 0$ and $a < 0$ separately.

【6】 (Hint) Directly calculate $\mathrm{d}\left(\int_{-\infty}^{\infty} f(t)\mathrm{e}^{-j\omega t}\mathrm{d}t\right)/\mathrm{d}\omega$.

【7】 (Hint) Repeat $\mathcal{F}\left[\dfrac{\mathrm{d}f(t)}{\mathrm{d}t}\right] = j\omega\mathcal{F}[f(t)]$.

【8】 A. $f(t) = \dfrac{1(1 + t) + 1(1 - t) - 1}{2}$. B. $\mathrm{e}^{-at}1(t)$.

【9】 【10】 【11】 【12】 Omitted.

Advanced Level

【1】 (Hint) Use $\mathcal{F}[1(t)] = \pi\delta(\omega) + \frac{1}{j\omega}$ and let $f(t) = f_r(t) + jf_i(t)$ and $F(\omega) = F_r(\omega) + jF_i(\omega)$. The inverse Fourier transform gives $f_r(t) = \frac{1}{2\pi}\int_{-\infty}^{\infty}[F_r(\omega)\cos\omega t - F_i(\omega)\sin\omega t]\mathrm{d}t$.

【2】 (Hint) Use the residue theorem to solve it.

【3】 【4】 【5】 Omitted.

★ Chapter 5

Basic Level

【1】 Refer to Section 5.2.

【2】 The highest angular frequency is $\omega_M = 30\pi$ rad/s. Let, $\omega_S = 2 \times \omega_M = 60\pi$ rad/s, that is, $\Delta T = \frac{2\pi}{\omega_S} = \frac{1}{30}$ s. Aliasing occurs if a sampling period is larger than ΔT.

【3】 (Hint) Refer to Section 5.1. And consider sampling a sine wave, $\sin\omega t$ at $\omega t = k\pi + \pi/2$ and $\omega t = k\pi$ $(k = 0, 1, 2, \ldots)$.

【4】 For $\Delta T = 2$ s, the oscillation is sampled only once per period. For $\Delta T = 1.8$ s, aliasing appears at the angular frequency $\frac{2\pi}{1.8 \times 2} - \left(\pi - \frac{2\pi}{1.8 \times 2}\right) = \frac{\pi}{9}$ rad/s.

【5】 Refer to the sampling theorem.

【6】 Omitted.

Advanced Level

【1】 Check the sample program `ad_prob_05_1.m`. Change the number of sample points (thus, the sampling period) in the program and examine the difference in results.

【2】 Let $y(t) = \mathrm{d}s(t)/\mathrm{d}t$. Using the differentiation property of the Fourier transform yields $Y(\omega) = j\omega S(\omega)$. The bandwidth of $s(t)$, ω_M, means that $S(\omega) = 0$, for $|\omega| > \omega_M$. That is also true for $\omega S(\omega)$. Thus, $y(t)$ has the same bandwidth as $s(t)$ does.

★ Chapter 6

Basic Level

【1】 A. $F_0 = N$ and $F_k = 0$ $(k = 1, 2, \ldots, N-1)$.

B. $F_0 = 0$ and $F_k = 1 - \mathrm{e}^{-j2k(N-1)\pi/N}$ $(k = 1, 2, \ldots, N-1)$.

【2】 $F_{N+k} = \displaystyle\sum_{m=0}^{N-1} f_m W_N^{m(N+k)} = \sum_{m=0}^{N-1} f_m W_N^{mN+mk} = \sum_{m=0}^{N-1} f_m W_N^{mk} = F_k.$

【3】 A. $[0, -2-j2, 0, -2+j2]^{\mathrm{T}}$. B. $[1.5, -0.5+j0.5, -0.5, -0.5-j0.5]^{\mathrm{T}}$.

C. $[0, 2, 0, 2]^{\mathrm{T}}$.

【4】 A. $[2.4142, -j2, -1, 0, -0.4142, 0, -1, j2]^{\mathrm{T}}$.

B. $[0.6875, 0.4268, 0.1250, 0.0732, 0.0625, 0.0732, 0.1250, 0.4268]^{\mathrm{T}}$.

C. $[3.9090, 0.7691 - j0.9426, 0.5382 - j0.4192, 0.4952 - j0.1758, 0.4861,$
$0.4952 + j0.1758, 0.5382 + j0.4192, 0.7691 + j0.9426]^{\mathrm{T}}$.

【5】 (Hint) Use (6.21).

【6】 (Hint) Use Figure 6.4.

【7】 (Hint) $\omega_M = 4\pi$ rad/s. ΔT should not be larger than 0.25 s.

【8】 (Hint) $\omega_M = \pi$ rad/s. ΔT should not be larger than 1 s.

Advanced Level

【1】 Omitted.

【2】 Main comments: `y=fft(s); yr=real(y); yi=imag(y);`.

【3】 Omitted.

【4】 Main comments: `L=length(y); wrec=rectwin(L); wham=Hamming(L);`
`yrec=y.*wrec; yham=y.*wham;`.

★ Chapter 7

Basic Level

【1】 【2】 【3】 【4】 Omitted.

【5】 $\displaystyle\lim_{T_a\to\infty} \frac{1}{T_a}\int_{-T_a/2}^{T_a/2} |f_T(t)|^2 \mathrm{d}t = \frac{1}{T}\int_{-T/2}^{T/2} |f_T(t)|^2 \mathrm{d}t.$

Let $P(\omega) = \dfrac{1}{T}\left|\displaystyle\int_{-T/2}^{T/2} f_T(t)\mathrm{e}^{-j\omega t}\mathrm{d}t\right|^2.$

$$\frac{1}{T}\int_{-T/2}^{T/2}|f_T(t)|^2\mathrm{d}t = \frac{1}{2\pi}\int_{-\infty}^{\infty}P(\omega)\mathrm{d}\omega$$
$$= \int_{-\infty}^{\infty}\left[\sum_{n=-\infty}^{\infty}|c_n|^2\delta(\omega - n\omega_0)\right]\mathrm{d}\omega$$
$$= \sum_{n=-\infty}^{\infty}|c_n|^2\int_{-\infty}^{\infty}\delta(\omega - n\omega_0)\mathrm{d}\omega$$
$$= \sum_{n=-\infty}^{\infty}|c_n|^2.$$

【6】 Omitted.

Advanced Level

【1】 【2】 【3】 Omitted.

【4】 See `07_ad_prob4.m`.

★ Chapter 8

Basic Level

【1】 A. (a) $v_R(t) = \mathcal{F}^{-1}\left(\dfrac{10}{10+j\omega}\right) = 10\,\mathrm{e}^{-10t}\times 1(t)$.

(b) $x(t) = \mathcal{F}^{-1}\left(\dfrac{1}{0.1-\omega^2}\right) = \sqrt{10}\sin\sqrt{0.1}t \times 1(t)$.

B. (a) $v_R(t) = \mathcal{F}^{-1}\left(\dfrac{10}{10+j\omega}\dfrac{1}{1+j\omega}\right) = \dfrac{10}{9}\left(\mathrm{e}^{-t} - \mathrm{e}^{-10t}\right)\times 1(t)$.

(b) $x(t) = \mathcal{F}^{-1}\left(\dfrac{1}{0.1-\omega^2}\dfrac{1}{1+j\omega}\right) = \dfrac{10}{11}\Big[\sqrt{10}(\sin\sqrt{0.1}t - \cos\sqrt{0.1}t)$
$+\,\mathrm{e}^{-t}\Big]\times 1(t)$.

【2】 $y(t) = \mathcal{F}^{-1}\left[G(\omega)R(\omega)\right] = \mathcal{F}^{-1}\left[K\mathrm{e}^{-j\omega t_0}R(\omega)\right]$

$$= \frac{1}{2\pi}\int_{-\infty}^{\infty}K\mathrm{e}^{-j\omega t_0}R(\omega)\mathrm{e}^{j\omega t}\mathrm{d}\omega$$
$$= \frac{K}{2\pi}\int_{-\infty}^{\infty}R(\omega)\mathrm{e}^{j\omega(t-t_0)}\mathrm{d}\omega$$
$$= Kr(t-t_0).$$

【3】 Step 1) Derive ordinary differential equations for the partial differential equation. Assuming that $u(x,y) = X(x)T(t)$ and substituting it into the equation in the problem yield $X(x)\dfrac{\mathrm{d}T(t)}{\mathrm{d}t} = T(t)\dfrac{\mathrm{d}^2X}{\mathrm{d}x^2}$. That is,

$\frac{\mathrm{d}T(t)}{\mathrm{d}t}/T(t) = \frac{\mathrm{d}^2X(x)}{\mathrm{d}x^2}/X(x)$. Note that the terms on the left and right sides of this equation depend only on t and x, respectively. Thus, both sides must be constant. Letting a be such a constant gives $\frac{\mathrm{d}T(t)}{\mathrm{d}t}/T(t) = \frac{\mathrm{d}^2X(x)}{\mathrm{d}x^2}/X(x) = a$. As a result, we obtain two ordinary differential equations

$$\begin{cases} \dfrac{\mathrm{d}^2X(x)}{\mathrm{d}x^2} - aX(x) = 0, \\ \dfrac{\mathrm{d}T(t)}{\mathrm{d}t} - aT(t) = 0. \end{cases} \tag{1}$$

Step 2) Find solutions satisfying boundary conditions. Omitting detailed discussion, we can see that there is no meaningful solution to the first differential equation in (1) for $a = \omega^2 \geqq 0$ that satisfies the boundary condition. Therefore, we let $\omega > 0$ and $a = -\omega^2 (< 0)$, and obtain

$$\begin{cases} \dfrac{\mathrm{d}^2X(x)}{\mathrm{d}x^2} + \omega^2 X(x) = 0, \\ \dfrac{\mathrm{d}T(t)}{\mathrm{d}t} - \omega^2 T(t) = 0. \end{cases} \tag{2}$$

The general solution to the first equation in (2) is given by

$$X(x) = X_1 \sin \omega x + X_2 \cos \omega x, \tag{3}$$

where X_1 and X_2 are constants. It follows from the boundary condition $X(0) = 0$ and $X(1) = 0$ that

$$\begin{cases} X(0) = X_2 = 0, \\ X(1) = X_1 \sin \omega + X_2 \cos \omega = X_1 \sin \omega = 0, \\ \qquad\qquad\qquad\qquad\qquad\qquad \omega = n\pi,\ n = 1, 2, \ldots \end{cases} \tag{4}$$

Thus, $X(x)$ is given by

$$X(x) = X_1 \sin n\omega x. \tag{5}$$

Next, find the solution to the second equation in (2)

$$T(t) = T_1 \mathrm{e}^{-\omega^2 t} = T_1 \mathrm{e}^{-n^2\pi^2 t}. \tag{6}$$

Combining (5) and (14) yields

$$u(x,t) = X_1T_1(\sin n\pi x)\mathrm{e}^{-n^2\pi^2 t}.$$

Redefining the linear combination coefficient as u_n $(n = 1, 2, 3, \ldots)$ gives

$$u(x,t) = \sum_{n=1}^{\infty} u_n(\sin n\pi x)\mathrm{e}^{-n^2\pi^2 t}, \ n = 1, 2, 3, \ldots. \tag{7}$$

Step 3) Find a unique solution for the entire problem. Substituting $t = 0$ into (7) yields

$$u(x,0) = \sum_{n=1}^{\infty} u_n \sin n\pi x.$$

The initial condition provides

$$u(x,0) = \begin{cases} 5, & \text{if } 0 < x \leq \dfrac{1}{2}, \\ 0, & \text{if } \dfrac{1}{2} < x < 1. \end{cases}$$

Choosing u_n such that $u(x,0)$ is the Fourier series of the constants 5 and 0, we have

$$u_n = 2\int_0^1 u(x,0)\sin n\pi x \mathrm{d}x = 2\int_0^{1/2} 5\sin n\pi x \mathrm{d}x = \frac{10}{n\pi}\left(1 - \cos\frac{n\pi}{2}\right).$$

Hence, the solution to the problem is

$$u(x,t) = \sum_{n=1}^{\infty} \frac{10}{n\pi}\left(1 - \cos\frac{n\pi}{2}\right)\sin n\pi x \times \mathrm{e}^{-n^2\pi^2 t}.$$

【4】 Step 1) Derive ordinary differential equations for the partial differential equation: Assuming that $u(x,y) = X(x)Y(y)$ and substituting it into the equation in the problem yields $\dfrac{\mathrm{d}^2X(x)}{\mathrm{d}x^2}Y(y) + \dfrac{\mathrm{d}^2Y(y)}{\mathrm{d}y^2}X(x) = 0$. That is, $-\dfrac{\mathrm{d}^2X(x)/\mathrm{d}x^2}{X(x)} = \dfrac{\mathrm{d}^2Y(y)/\mathrm{d}y^2}{Y(y)}$. Note that the terms on the left and right sides of this equation depend only on x and y, respectively. Thus, both sides must be constant. Letting a be such a constant gives $-\dfrac{\mathrm{d}^2X(x)/\mathrm{d}x^2}{X(x)} = \dfrac{\mathrm{d}^2Y(y)/\mathrm{d}y^2}{Y(y)} = a$. As a result, we obtain two ordinary differential equations

$$\begin{cases} \dfrac{\mathrm{d}^2X(x)}{\mathrm{d}x^2} + aX(x) = 0, \\ \dfrac{\mathrm{d}^2Y(y)}{\mathrm{d}y^2} - aY(y) = 0. \end{cases} \tag{8}$$

Step 2) Find solutions satisfying boundary conditions. Similar to the answer to Basic-Level Problem 3, we let $a = -\omega^2 < 0$ ($\omega > 0$) and obtain

$$\begin{cases} \dfrac{\mathrm{d}^2X(x)}{\mathrm{d}x^2} - \omega^2 X(x) = 0, \\ \dfrac{\mathrm{d}^2Y(y)}{\mathrm{d}y^2} + \omega^2 Y(y) = 0. \end{cases} \tag{9}$$

Using the linear combination provides the following fundamental solution to the first equation in (9)

$$X(x) = X_1\mathrm{e}^{n\pi x} + X_2\mathrm{e}^{-n\pi x}, \tag{10}$$

where X_1 and X_2 are constants. It follows from the boundary condition $u(1, y) = 0$ that $X_1\mathrm{e}^{n\pi} + X_2\mathrm{e}^{-n\pi} = 0$, or

$$X_2 = -X_1\mathrm{e}^{2n\pi} = 0. \tag{11}$$

Thus, $X(x)$ is

$$X(x) = X_1\mathrm{e}^{n\pi}\left[\mathrm{e}^{n\pi(x-1)} - \mathrm{e}^{-n\pi(x-1)}\right]. \tag{12}$$

Since $\sinh x = (\mathrm{e}^x - \mathrm{e}^{-x})/2$, (12) is

$$X(x) = 2X_1\mathrm{e}^{n\pi}\sinh n\pi(x-1). \tag{13}$$

Next, find the solution to the second equation in (9)

$$Y(y) = Y_1(t)\sin\omega y + Y_2(t)\cos\omega y, \tag{14}$$

where Y_1 and Y_2 are constants. The boundary conditions $u(x, 0) = 0$ and $u(x, 1) = 0$ yield $Y(0) = Y(1) = 0$. Thus,

$$Y_2 = 0,\ Y_1\sin\omega = 0,\ \omega = n\pi,\ n = 1, 2, 3, \ldots.$$

As a result, the general solution is

$$Y(y) = Y_1\sin n\pi y,\ \omega = n\pi,\ n = 1, 2, 3, \ldots. \tag{15}$$

Multiplying the two solutions $X(x)$ and $Y(y)$ provides us with

$$u(x,y) = 2X_1 e^{n\pi} \sinh n\pi(x-1) \times Y_1 \sin ny.$$

Redefining the linear combination coefficient as u_n $(n = 1, 2, 3, \ldots)$ gives

$$u(x,y) = \sum_{n=1}^{\infty} u_n \sinh n\pi(x-1) \sin n\pi y, \; n = 1, 2, 3, \ldots. \qquad (16)$$

Step 3) Find a unique solution for the entire problem. Substituting the boundary condition $u(0,y) = 1$ into (16) yields

$$u(0,y) = \sum_{n=1}^{\infty} u_n \sinh(-n\pi) \sin n\pi y = 1.$$

Letting $v_n = u_n \sinh(-n\pi)$, the initial condition becomes

$$u(0,y) = \sum_{n=1}^{\infty} v_n \sin n\pi y = 1.$$

Choosing u_n such that $u(0,y)$ is the Fourier series of the constant 1, we have

$$v_n = 2\int_0^1 \sin n\pi y \mathrm{d}y = 2\left[-\frac{1}{n\pi}\cos n\pi y\right]_0^1 = \frac{2}{n\pi}(1-\cos n\pi).$$

Hence, $u_n = \dfrac{2(1-\cos n\pi)}{n\pi \sinh(-n\pi)}$ and the solution to the problem is

$$u(x,y) = \sum_{n=1}^{\infty} \frac{2(1-\cos n\pi)}{n\pi \sinh(-n\pi)} \sinh n\pi(x-1) \sin n\pi y.$$

Advanced Level

【1】 A. (a) $v_R(t) = \mathcal{F}^{-1}\left[\dfrac{j\omega}{(j\omega)^2 + j\omega + 1}\left(\pi\delta(\omega) + \dfrac{1}{j\omega}\right)\right] = \dfrac{2}{\sqrt{3}} e^{-0.5t} \sin\dfrac{\sqrt{3}}{2}t$ $\times 1(t)$ and

(b) $x(t) = \mathcal{F}^{-1}\left[\dfrac{j\omega}{(j\omega)^2 + j\sqrt{2}\omega + 1}\left(\pi\delta(\omega) + \dfrac{1}{j\omega}\right)\right]$

$= \left[1 - \sqrt{2} e^{-\sqrt{2}t/2} \sin\left(\dfrac{\sqrt{2}}{2}t - \dfrac{\pi}{4}\right)\right] \times 1(t)$. Note that $\omega\delta(\omega) = 0$ [see (2.49)].

B. (a) $v_R(t) = \mathcal{F}^{-1}\left[\dfrac{j\omega}{(j\omega)^2 + j\omega + 1} \dfrac{1}{(j\omega)^2 + 1}\right]$

$$= \left(\sin t - \frac{2\sqrt{3}}{3}\mathrm{e}^{-0.5t}\sin\frac{\sqrt{3}}{2}t\right) \times 1(t) \text{ and}$$

(b) $x(t) = \mathcal{F}^{-1}\left[\dfrac{j\omega}{(j\omega)^2 + j\sqrt{2}\omega + 1}\dfrac{1}{(j\omega)^2+1}\right]$

$$= \left(\frac{\sqrt{2}}{2}\sin t - \mathrm{e}^{-\sqrt{2}t/2}\sin\frac{\sqrt{2}}{2}t\right) \times 1(t).$$

【2】 A. (Hint) Calculate the output for an input $v_i(t) = V_0 + \sum_{n=1}^{\infty} V_n \cos n\omega t$ and use the additivity property to calculate all the output.

B. $P = \dfrac{1}{T}\displaystyle\int_{-T/2}^{T/2} v_i(t)i(t)\mathrm{d}t$

$$= \frac{1}{T}\int_{-T/2}^{T/2}\left[V_0 + \sum_{n=1}^{\infty} V_n\cos n\omega t\right]\left[I_0 + \sum_{m=1}^{\infty} I_m\cos(m\omega t + \phi_m)\right]\mathrm{d}t$$

$$= \frac{1}{T}\int_{-T/2}^{T/2}\left[V_0I_0 + V_0\sum_{m=1}^{\infty} I_m\cos(m\omega t + \phi_m) + I_0\sum_{n=1}^{\infty} V_n\cos n\omega t\right.$$

$$\left.+\sum_{n=1}^{\infty}\sum_{m=1}^{\infty} V_nI_m\cos n\omega t\cos(m\omega t + \phi_m)\right]\mathrm{d}t$$

$$= V_0I_0 + \frac{1}{2}\sum_{n=1}^{\infty} V_nI_n\cos\phi_n.$$

【3】 Step 1) Derive ordinary differential equations for the partial differential equation: Assuming that $u(x,y,t) = T(t)U(x,y)$ and substituting it into (8.50) yield $\dfrac{\mathrm{d}^2T(t)}{\mathrm{d}t^2}U(x,y) = c^2T(t)\left[\dfrac{\partial^2U(x,y)}{\partial x^2} + \dfrac{\partial^2U(x,y)}{\partial y^2}\right]$. Thus, $\dfrac{\mathrm{d}^2T(t)/\mathrm{d}t^2}{c^2T(t)} = \dfrac{1}{U(x,y)}\left[\dfrac{\partial^2U(x,y)}{\partial x^2} + \dfrac{\partial^2U(x,y)}{\partial y^2}\right]$. Note that the terms on the left and right sides of the above equation depend only on t, and x and y, respectively. Thus, both sides must be constant. So, letting k be such a constant yields

$$\begin{cases} \dfrac{\partial^2U(x,y)}{\partial x^2} + \dfrac{\partial^2U(x,y)}{\partial y^2} - kU(x,y) = 0, \\ \dfrac{\mathrm{d}^2T(t)}{\mathrm{d}t^2} - c^2kT(t) = 0. \end{cases} \tag{17}$$

We further assume that $U(x,y) = X(x)Y(y)$ and obtain $\dfrac{\mathrm{d}^2X(x)/\mathrm{d}x^2}{X(x)} = -\dfrac{\mathrm{d}^2Y(y)/\mathrm{d}y^2}{Y(y)} + k$. Since the terms on the left and right sides of the above equation depend only on x and y, both sides must be constant. Letting the constant be $-\alpha$ and defining $\beta = -(\alpha + k)$ yield $\dfrac{\mathrm{d}^2 X(x)}{\mathrm{d}x^2} + \alpha X(x) = 0$ and $\dfrac{\mathrm{d}^2 Y(y)}{\mathrm{d}y^2} + \beta Y(y) = 0$.

Step 2) Find the solutions satisfying boundary conditions: The boundary conditions $u(0, y, t) = u(a, y, t) = u(x, 0, t) = u(x, b, t) = 0$ and $t \geqq 0$ provide $X(0) = X(a) = 0$ and $Y(0) = Y(b) = 0$. Thus, there is no meaningful solution for $\alpha \leqq 0$ and $\beta \leqq 0$. Letting $\alpha > 0$ and $\beta > 0$ yields $X(x) = A\cos\sqrt{\alpha}x + B\sin\sqrt{\alpha}x$ and $Y(y) = C\cos\sqrt{\beta}y + D\sin\sqrt{\beta}y$. It follows from the boundary conditions that $A = 0$, $B\sin\sqrt{\alpha}a = 0$, $C = 0$, and $D\sin\sqrt{\beta}b = 0$. $B\sin\sqrt{\alpha}a = 0$ and $D\sin\sqrt{\beta}b = 0$ provide $\alpha = \left(\dfrac{m\pi}{a}\right)^2$ and $\beta = \left(\dfrac{n\pi}{b}\right)^2$ $(m, n = 1, 2, \ldots)$. Hence, the general solution $U(x, y)$ of this ordinary differential equation is $U_{mn}(x, y) = \sin\dfrac{m\pi x}{a}\sin\dfrac{n\pi y}{b}$ and $k_{mn} = -\alpha - \beta = -\left(\dfrac{m\pi}{a}\right)^2 - \left(\dfrac{n\pi}{b}\right)^2$ $(m, n = 1, 2, \ldots)$. Substituting the above results into the second equation in (17) yields $\dfrac{\mathrm{d}^2 T(t)}{\mathrm{d}t^2} + \omega_{mn}^2 T(t) = 0$ and $\omega_{mn} = c\sqrt{\left(\dfrac{m\pi}{a}\right)^2 + \left(\dfrac{n\pi}{b}\right)^2}$ $(m, n = 1, 2, \ldots)$. The general solution of the above differential equation is $T_{mn}(t) = A_{mn}\cos\omega_{mn}t + B_{mn}\sin\omega_{mn}t$. This gives $u_{mn}(x, y, t) = (A_{mn}\cos\omega_{mn}t + B_{mn}\sin\omega_{mn}t)\sin\dfrac{m\pi x}{a}\sin\dfrac{n\pi y}{b}$ $(m, n = 1, 2, \ldots)$.

Step 3) Find a solution for the entire problem: The principle of superposition provides $u(x, y, t) = \displaystyle\sum_{m=1}^{\infty}\sum_{n=1}^{\infty} u_{mn}(x, y, t)$

$= \displaystyle\sum_{m=1}^{\infty}\sum_{n=1}^{\infty}(A_{mn}\cos\omega_{mn}t + B_{mn}\sin\omega_{mn}t)\sin\frac{m\pi x}{a}\sin\frac{n\pi y}{b}$. The solution of $u(x, y, t)$ that satisfies the initial condition at the $t = 0$ is

$$\left\{\begin{aligned} & u(x, y, 0) = \sum_{m=1}^{\infty}\sum_{n=1}^{\infty} A_{mn}\sin\frac{m\pi x}{a}\sin\frac{n\pi y}{b} = f(x, y), \\ & \left.\frac{\partial u(x, y, t)}{\partial t}\right|_{t=0} = \sum_{m=1}^{\infty}\sum_{n=1}^{\infty} B_{mn}\omega_{mn}\sin\frac{m\pi x}{a}\sin\frac{n\pi y}{b} \\ & \qquad = g(x, y). \end{aligned}\right. \tag{18}$$

Equations in (18) are called double Fourier series. Then, we find the coefficients A_{mn} and B_{mn}. For this purpose, we define $C_m(y)$

$= \sum_{n=1}^{\infty} A_{mn} \sin \frac{n\pi y}{b}$ and rewriting the above equations yields $f(x,y) = \sum_{m=1}^{\infty} C_m(y) \sin \frac{m\pi x}{a}$. For a fixed y, $f(x,y) = \sum_{m=1}^{\infty} C_m(y) \sin \frac{m\pi x}{a}$ is a Fourie series of $f(x,y)$. Since $f(x,y)$ is odd with respect to both x and y, $C_m(y) = \frac{2}{a}\int_0^a f(x,y) \sin \frac{m\pi x}{a} \mathrm{d}x$. Note that $C_m(y) = \sum_{n=1}^{\infty} A_{mn} \sin \frac{n\pi y}{b}$.
A_{mn} is given by
$A_{mn} = \frac{2}{b}\int_0^b C_m(y) \sin \frac{n\pi y}{b} \mathrm{d}y$. Thus,
$A_{mn} = \frac{4}{ab}\int_0^b \int_0^a f(x,y) \sin \frac{m\pi x}{a} \sin \frac{n\pi y}{b} \mathrm{d}x\mathrm{d}y$. In the same manner, we obtain
$B_{mn} = \frac{4}{ab\omega_{mn}}\int_0^b \int_0^a g(x,y) \sin \frac{m\pi x}{a} \sin \frac{n\pi y}{b} \mathrm{d}x\mathrm{d}y$.
Now, we obtained a solution for the given boundary and initial conditions.

【4】 Omitted.

★ Chapter 9

Basic Level

【1】 A. $F(\omega_1, \omega_2) = \frac{1}{2\pi}$.

B. $F(\omega_1, \omega_2) = 2\pi^2[\delta(\omega_1 - 2\pi a, \omega_2 - 2\pi b) + \delta(\omega_1 + 2\pi a, \omega_2 + 2\pi b)]$.

C. $F(\omega_1, \omega_2) = \mathrm{e}^{-(\omega_1^2+\omega_2^2)/(4\pi)}$.

【2】 $[a_0, a_1, a_2] = [0.80, 0.23, 0]$.

【3】 Omitted.

【4】 $\{a, b, c, d\} = \{3, 5, 7, 9\}$.

【5】 Image compression is carried out by the following steps: First, perform the DST of an image and decompose it into its spatial frequency components. Then, delete the high-frequency components. Finally, perform the IDFT.

Advanced Level

【1】 【2】 Omitted.

★ Chapter 10

Basic Level

【1】 A. $\frac{1}{s}$. B. $\frac{s}{s^2+a^2}$. C. $\frac{a}{s^2+a^2}$. D. $s^2Y(s)$.

【2】 A. $1-e^{-t}$. B. $\frac{2}{\sqrt{3}}e^{-t/2}\sin\frac{\sqrt{3}}{2}t$.

C. $2e^{-3t}-e^{-2t}$. D. e^{at}. E. $\cos 3t$.

【3】 A. $1-e^{-t/2}$. B. $e^{-2t}(\cos t+\sin t)$.

Advanced Level

【1】 Using the convolution property of Laplace Transform yields

A. $\mathcal{L}\left[\int_0^t \sin(t-x)\cos x\mathrm{d}x\right] = \mathcal{L}(\sin t)\mathcal{L}(\cos t) = \frac{1}{s^2+1}\frac{s}{s^2+1} = \frac{s}{(s^2+1)^2}$.

B. $\mathcal{L}\left[\int_0^t (1-x)e^{a(t-x)}\mathrm{d}x\right] = \mathcal{L}(1-t)\mathcal{L}(e^{at}) = \left(\frac{1}{s}-\frac{1}{s^2}\right)\frac{1}{s-a} = \frac{s-1}{s^2(s-a)}$.

【2】 A. $\frac{1}{4}+\frac{1}{4}\cos 2t$ and B. $\frac{1}{4}+\frac{1}{12}e^{-4t}-\frac{1}{3}e^{-t}$. The DC components of both are $\frac{1}{4}$. In the steady state, while the response of A. oscillates at an angular frequency 2 rad/s, that of B. exponentially converges to 0.

【3】 Applying the Laplace transform for the independent variable t to the one-dimensional wave equation, $\mathcal{L}\left[\frac{\partial^2 u(x,t)}{\partial t^2}\right] = c^2\mathcal{L}\left[\frac{\partial^2 u(x,t)}{\partial x^2}\right]$, yields $s^2\mathcal{L}\left[u(x,t)\right]-su(x,0)-\left.\frac{\partial u(x,t)}{\partial t}\right|_{t=0} = c^2\mathcal{L}\left[\frac{\partial^2 u(x,t)}{\partial x^2}\right]$. So, $s^2\mathcal{L}\left[u(x,t)\right] = c^2\mathcal{L}\left[\frac{\partial^2 u(x,t)}{\partial x^2}\right]$. Assuming that differential and integral operations are interchangeable, we have $\mathcal{L}\left[\frac{\partial^2 u(x,t)}{\partial x^2}\right] = \int_0^\infty e^{-st}\frac{\partial^2 u(x,t)}{\partial x^2}\mathrm{d}t$ $= \frac{\partial^2}{\partial x^2}\int_0^\infty e^{-st}u(x,t)\mathrm{d}t = \frac{\partial^2}{\partial x^2}\mathcal{L}\left[u(x,t)\right]$. Defining $U(x,s) = \mathcal{L}[u(x,t)]$ yields $s^2U = c^2\frac{\partial^2 U}{\partial x^2}$. Hence, $\frac{\partial^2 U}{\partial x^2} - \frac{s^2}{c^2}U = 0$. Solving it for an independent variable x yields $U(x,s) = A(s)e^{sx/c}+B(s)e^{-sx/c}$. Assuming that the limit and integral operations are interchangeable, we obtain $\lim_{x\to\infty}U(x,s) = \lim_{x\to\infty}\int_0^\infty e^{-st}u(x,t)\mathrm{d}t = \int_0^\infty e^{-st}\lim_{x\to\infty}u(x,t)\mathrm{d}t = 0$. So, the term $A(s)e^{sx/c}$ must be zero. And $\lim_{x\to\infty} |A(s)e^{sx/c}| = \infty$ when $c > 0$, Hence, $A(s) = 0$. It is easy to find that $U(0, s) = B(s) = \mathcal{L}[f(t)] = F(s)$. As a result, $U(x, s) = F(s)e^{-sx/c}$. Applying the inverse Laplace transform for the above equation yields $u(x, t) = f\left(t - \frac{x}{c}\right) 1\left(t - \frac{x}{c}\right)$. As an example, let $f(t) = \begin{cases} \sin t, & \text{if } 0 \leqq t \leqq 2\pi, \\ 0, & \text{if } t < 0 \text{ or } t > 2\pi, \end{cases}$ then

$$u(x, t) = \begin{cases} \sin\left(t - \frac{x}{c}\right), & \text{if } \frac{x}{c} \leqq t \leqq \frac{x}{c} + 2\pi, \\ 0, & \text{if } t < \frac{x}{c} \text{ or } t > \frac{x}{c} + 2\pi. \end{cases}$$

This solution shows that an isolated sine wave with its length of 2π propagates along the x -axis with a speed c.

Index

―― 著者略歴 ――

佘　錦華（しゃ　きんか）
1993年　東京工業大学大学院理工学研究科博士課程修了（制御工学専攻），博士（工学）
1993年　東京工科大学講師
2001年　東京工科大学助教授
2007年　東京工科大学准教授
2010年　東京工科大学教授
現在に至る

宮本　皓（みやもと　こう）
2014年　筑波大学理工学群社会工学類都市計画主専攻卒業
2016年　東京工業大学大学院総合理工学研究科修士課程修了（環境理工学創造専攻）
2018年　日本学術振興会特別研究員（DC2）
2019年　東京工業大学環境・社会理工学院博士課程修了（建築学系），博士（工学）
2019年　清水建設株式会社技術研究所勤務
現在に至る

川田　誠一（かわた　せいいち）
1977年　大阪大学工学部産業機械工学科卒業
1979年　大阪大学大学院工学研究科博士前期課程修了（機械工学専攻）
1982年　大阪大学大学院工学研究科博士後期課程単位取得退学（機械工学専攻）
1982年　大阪大学助手
1983年　工学博士（大阪大学）
1986年　東京都立大学助手
1990年　東京都立大学助教授
2000年　東京都立大学教授
2005年　首都大学東京教授
2006年　東京都立産業技術大学院大学教授
2022年　東京都立産業技術大学院大学名誉学長・名誉教授
2022年　中国地質大学（武漢）教授
現在に至る

英語で学ぶ　**フーリエ解析とその応用**
Introduction to Fourier Analysis and Its Applications

2023 年 5 月 11 日　初版第 1 刷発行 ★

検印省略

著　者	佘　錦　華
	宮　本　皓
	川　田　誠　一
発 行 者	株式会社　コ ロ ナ 社
	代 表 者　牛 来 真 也
印 刷 所	三 美 印 刷 株 式 会 社
製 本 所	有限会社　愛 千 製 本 所

112–0011　東京都文京区千石 4–46–10
発 行 所　株式会社　コ ロ ナ 社
CORONA PUBLISHING CO., LTD.
Tokyo Japan
振替 00140–8–14844・電話(03)3941–3131(代)
ホームページ https://www.coronasha.co.jp

ISBN 978–4–339–06127–7　C3041　Printed in Japan　(新宅)